EBURY PRESS

I'M A CLIMATE OPTIMIST

Aakash Ranison has been listed under *Forbes India's* Top 100 Digital Stars 2022, awarded Climate Warrior of the Year 2022 and 2023 (editor's choice) by *Cosmopolitan* magazine, and Sustainable Changemaker of the Year 2022 (popular choice) by *Exhibit*. He has delivered five TEDx Talks, authored his first book on climate change in 2020 and created three climate art installations, which have been covered by over 230 national and international media publications. He is also a contributing author for *Forbes India*, *Times of India* and YourStory.

On the road for eight years, Aakash lives a vegan, nomadic and minimalist life and spends his time traveling the length and breadth of India by walking, cycling, hitchhiking, or public transport with his trusty backpack and camera, spreading his message of love, kindness and sustainability.

For more information visit: ***aakashranison.com/climate-optimist***
Get in touch with the author: namaste@aakashranison.com
Instagram: instagram.com/aakashranison
Twitter: twitter.com/aakashranison
Youtube: youtube.com/aakashranison
Linkedin: linkedin.com/in/aakasl

Celebrating 35 Years of
Penguin Random House India

INDIA'S FIRST CARBON NEUTRAL BOOK

I'M A CLIMATE OPTIMIST

An easy guide to lead a sustainable life.

AAKASH RANISON

Awarded **Climate Warrior 2022 & 2023** by *Cosmopolitan*

EBURY
PRESS

An imprint of Penguin Random House

EBURY PRESS

USA | Canada | UK | Ireland | Australia
New Zealand | India | South Africa | China

Ebury Press is part of the Penguin Random House group of companies
whose addresses can be found at global.penguinrandomhouse.com

Published by Penguin Random House India Pvt. Ltd
4th Floor, Capital Tower 1, MG Road,
Gurugram 122 002, Haryana, India

First published in Ebury Press by Penguin Random House India 2023

ISBN 9780143460428

Typeset in Gotham by Manipal Technologies Limited, Manipal

www.penguin.co.in

I dedicate this book with all my love.

To my amazing mom, Rani, thank you for giving me my name, for always tolerating my nonsense, and for working from 9 a.m. to 10 p.m. every day for years to pay my school fee even though I failed in almost all the classes. (School wasn't of much use, though. And, I never ended up going to a college).

*

To my life teachers, most of whom I have never met but whose work gave me a sense of life. They are my real teachers: Swami Vivekananda, Shaheed Bhagat Singh, His Holiness the 14th Dalai Lama, Thích Nhất Hạnh, Lao Tzu, Confucius, Piyush Pandey, Naval Ravikant, Simon Sinek, and Peter McKinnon.

*

And to my sunflower, Purvi, thank you for being my driving force.
Thank you for being the awesomeness that you are.

*

She is the one rising from the east.
She is the mother of the universe on the feast.
She is the creator of time and destroyer of the beast.
She is the Maa Kaali and I'm the priest.

Contents

1

Introduction: I'm a Climate Optimist

Hi! My name is Aakash Ranison. You might know me as an environmentalist, an advocate for sustainability, a nomadic traveller, a filmmaker, a friend or even an acquaintance.

Perhaps you don't know me at all—and that's just fine too.

My story is not a special one. It does not have dramatic turns of phrase, spine-chilling adventures, romantic narratives or even wry humour.

My story is not unique—because my story is your story. My story is the story of eight billion people around the world who are human beings just like me.

My story is our story.

It is the story of our collective existence on this planet, and the story of our planet itself.

But what is that story?

Today, conversations about climate change and conservation are marked by an air of fear, doom-signalling and cynicism. To put it simply, the story is this—our home is burning, and we are not doing enough to stop it.

Rapid climate change is threatening life as we know it, and if our current projections are right, we are unlikely to curtail our emissions before reaching that crucial tipping point between 1.2-1.5 degrees of atmospheric warming.

This tipping point symbolizes the collapse of our societal structures, the end of our civilization and the destruction of our natural ecosystems.

All around the world, scientists and younger generations are urging the powers that be to act to stop the destruction of all that we know. However, from the landmark Paris Agreement to a plethora of national-level legislative initiatives, it seems that the political powers at the helm of our world cannot commit meaningfully to enacting change. The rich keep stoking the fire and the poor suffer the stifling heat of the rising flames.

A study from January 2021 by the Research Institute for Humanity and Nature declared that the top 20 per cent of High Net Worth (HNW) households in India generate seven times the carbon emissions of low-expenditure households.[1] These marginalized socio-economic communities contribute little to the problem but bear the brunt of its weight.

Climate change sees no borders and no differences in language or GDP. It does not discriminate between the oppressor and the oppressed, the rich and the poor. Climate change is affecting us all.

Joining hands with the legislators of the world, the fossil fuel industry, animal agriculture institutions and the capitalistic structures that power the global market work in defiance of the actions taken to mitigate rapid climate change. Unable to overcome their fixation with short-term gains and their insatiable greed for more, they trade our future for a pretty penny, and have been doing so for years.

Piyush Pandey, chief creative officer worldwide and executive chairman India of Ogilvy & Mather and Padma Shri awardee, says:

'The number one thing about advertising is to understand your audience, not feed them. You have to understand, and then try to guide them towards what you think is a good way to live a life. The audience will consider what you say, and may choose to follow—this is their prerogative—but they will always consider your words. The essence of effective communication is respect for your audience. If you want to do meaningful business, if you want your audience to stay with you, you have a responsibility to be respectful and communicate honestly, and transparently, to foster a connection. The core of sustainability is about how long you can maintain this connection. The moment you are able to admit that you are not smarter than the consumer or the supplier, the moment you treat them with respect and guide them towards living a good way of life, you will form a connection that lasts forever. It

is not about knowing it all, but attempting to learn—
it is this attempt that makes things last. Brands
that greenwash are aware that they are lying, that
is why they wash their actions away with the story
they know they should have told. These brands will
never form a lasting connection. No one wants to
live with lies or liars. The Indian audience is smart.
We come from thousands of years of respect for the
environment and the world around us. We are not
here to be fooled, we are here to be conversed with,
and connected with.'

In these dire times, it is only natural that more and
more young people worldwide have been overcome by
anxious and depressive thoughts—a newly acknowledged
mental health crisis known as eco-anxiety.

Defined by the American Psychology Association
(APA) as 'the chronic fear of environmental cataclysm
that comes from observing the seemingly irrevocable
impact of climate change and the associated concern
for one's future and that of next generations', eco-
anxiety is plaguing youth populations across the world.[2]

A study that surveyed over 10,000 people aged
sixteen to twenty-five, from ten countries around the
world, found that 59 per cent were very or extremely
worried, while 84 per cent were at least moderately
worried about their future on this planet. Moreover,
over 50 per cent reported experiencing feelings
of sadness, anxiety, anger, helplessness and even
guilt. Perhaps most importantly, 39 per cent of the
participants were hesitant to have children in a world
with an uncertain future.

Percentage (%) of whole sample, and by country of negative thoughts about climate change and beliefs about government response to climate change. (n=1000 per country)

Thoughts about climate change	All Countries	Australia	Brazil	France	Finland	India	Nigeria	Phillipines	Portugal	UK	USA
People have failed to care for planet	83	81	92	77	75	86	76	93	89	80	78
Future is frightening	75	76	86	74	56	80	70	92	81	72	68
Humanity is doomed	56	50	67	48	43	74	42	73	62	51	46
Less opportunity than parents	55	57	50	61	42	67	49	70	54	53	44
Most valued will be destroyed	55	52	64	45	43	69	54	74	59	47	42
Family security will be threatened	52	48	65	50	30	65	55	77	52	39	35
Hesitant to have children	39	43	48	37	42	41	23	47	37	38	36

Table 1-1: Percentage of whole sample and by country of negative thoughts about climate change and beliefs about government response to climate change. (n = 1000 per country) Source: https://ssm.com/abstract=3918955

Percentage (%) of whole sample, and by country of beliefs about government response about climate change (n=1000 per country)

Thoughts about government response	All Countries	Australia	Brazil	France	Finland	India	Nigeria	Phillipines	Portugal	UK	USA
Failing young people	65	67	79	55	47	71	64	68	69	65	63
Lying about impact of actions taken	64	66	78	58	54	67	66	69	62	61	62
Dismissing peoples' distress	60	64	80	57	48	59	58	53	65	58	59
Betraying the future generations	58	59	77	49	46	66	55	56	62	57	56
Acting in line with climate science	36	33	22	28	38	53	40	52	38	32	28
Protecting me, planet and future gens	33	31	18	27	34	49	35	47	33	31	25
Can be trusted	31	30	22	23	34	51	31	40	32	28	21
Doing enough to avoid catastrophe	31	31	20	26	30	44	36	42	28	26	24

Table 1-2: Percentage of whole people, and by country of beliefs about government response about climate change. Source: https://ssm.com/abstract=3918955

These young people, with their whole lives in front of them, sit paralysed by fear and are worried about whether they will even have a future worth living for. In this age of uncertainty, the only course of action for most young people is to simply give up—after all, the world will end whether they try to save it or not.

However, this is simply not true.

Our environment is not doomed.

Our planet is not doomed.

We are not doomed.

The Blind Leading the Blind: A Lack of Vision

What we tend to forget while discussing climate change is the simple fact that our planet doesn't need our help. Our planet doesn't need us at all, and would probably prefer to be left alone.

The natural world is no stranger to extinction-level events. From the meteors that destroyed the dinosaurs to ice ages that lasted over 1,00,000 years, our planet has seen many species emerge, thrive and become lost to the immense power of nature.

Humankind has lived through the longest summer ever recorded on our planet, but just like the dinosaurs and all those who came and went after, there is no guarantee of survival in the future.

While our planet will survive the next extinction-level event, and the ones to come after, we very likely will not.

Most of us today are content to go about our lives, droning on through the mundane motions of schooling, building careers, finding a suitable mate, raising a family and dying in our sleep, all while exploiting our natural world and driving the knife deeper into its wounds, barely sparing a sideways glance at the destruction as we do so.

As we die, we leave the world a little worse for us having been there. With every birth, the cycle repeats itself. This is because our generation, like the generations before us, suffers from a lack of vision.

We live in a world of instant gratification—instant deliveries, instant messaging, instant noodles and even instant banking. We have lost our ability to be patient, work hard and see results and invest today in a future that will come to pass tomorrow.

But there are no quick fixes to climate change and global warming.

You see, it's not about *eradicating* carbon dioxide, methane or nitrous oxide—these are the ingredients that *make life possible*. The greenhouse effect is not a force for evil, but it is what *makes our atmosphere hospitable for life*. It is a state of imbalance that can turn these life-giving forces against us.

But there are no quick fixes to attaining balance, either. We need to realize that even if every person on the planet began fixing their ways now, there would still be a period of transition as our natural world corrects itself.

Think of our consumption patterns as a credit card—if your credit limit is a lakh, and you spend it but replenish it with a lakh of your money, your spending is balanced, and you are free from the danger of debt. However, if you spend two lakh, and do not have the money to pay the bank to offset the debt you have accumulated, you are in for a world of trouble and will have to work to pay it off.

In much the same way, we must offset our debts to our planet, and give back what we have taken from our natural world to keep it from turning against us.

But how do we do this?

To save ourselves, we must invest in the future. We must take advantage of the power of knowledge as it can instil trust in our individual choices. We must choose to think bigger, envision sustainable futures and work towards them patiently.

How Did We Get Here?

Sustainability has always been a core component of Indian culture. From the north to the south, ancient Indian philosophy and tradition have always underscored a sustainable way of life.

	NET Any restrictions on meat	Vegetarian	Non Vegetarian			
			Abstain from eating meat on certain days AND from eating certain meats	ONLY abstain from eating meat on certain days	ONLY abstain from eating certain meats	NO restrictions on meats
General population	81%	39%	30%	6%	5%	18%
Hindus	83	44	29	6	4	16
Muslims	67	8	39	7	14	32
Christians	66	10	30	10	16	33
Sikhs	82	59	16	3	4	16
Buddhists	84	25	38	14	8	15
Jains	97	92	4	1	0	3

Note: Don't know/Refused responses to any of the three questions included here are not shown. Figures may not add to subtotals indicated due to rounding. Respondents were asked if they are vegetarian. Those who said no were then asked two separate questions: "Do you abstain from eating meat on certain days?" and "Do you abstain from eating certain meats?"

Table 1-3: Majorities in all of India's religious groups follow at least some restrictions on meat in their diet. Source: Pew Research Centre[3]

For example, the yogic practice of *aparigraha*, or non-possession, which is a core principle of Jainism, upholds a minimalistic way of life, that is, keeping and consuming only what is necessary.

Even animal agriculture, one of the biggest contributors to global warming, has its solutions embedded in the whirlpool of Indian belief systems.

The vast melting pot of Indian cultures and traditions has aided biodiversity conservation efforts for millennia, with a 2017 Greendex report[4] ranking India as one of the few countries with a population that is conscious about its environmental footprint and is making the most sustainable choices.

Other uniquely Indian cultural phenomena, such as hoarding (saving seemingly useless items for future use) and thriftiness (the practice of reusing and repurposing) also contribute greatly to India's reputation as a sustainable and conscious nation. Think of your beloved old T-shirts that are now used

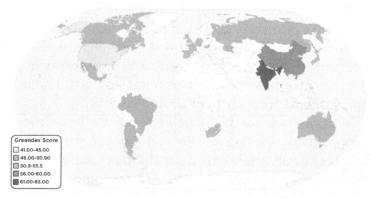

Figure 1-1: Greendex Index by country. Source: National Geographic/ GlobeScan Consumer Greendex[5]

as mops and dusters—this is a sustainable choice for our environment.

In fact, in 2011, over 70 per cent of India's population lived in sustainable rural communities, in harmony with nature and its flora and fauna.

However, as our economy develops and our infrastructural capacity grows, socio-economic trends are morphing to suit a new era of development. Our sustainable past in no way counterbalances the multitude of challenges that await us in the near future.

As I wrote this book, I encountered a problem I had always predicted would be the biggest challenge. The challenge was this: when we study the Western world, there is no lack of data to be found on any number of subjects. In nations like India, on the other hand, data is scattered, muddled or missing.

This presents a dire problem.

Data, due of its objectivity, has the power to separate lies from the truth. If you are a powerful multinational corporation, however, you can choose to buy, erase or modify the truth to suit your needs.

When the powers that be do not align with our interests or vision for the future, and hold the keys to the many truths so many of us miss, we can't help but be overcome by a sense of powerlessness.

But we are far from powerless. We have more power than we know. We can choose to be the difference. If there is one thing I learnt while researching this book, it is that we need more people actively investing in research, study and innovation to save our planet.

As a nation, and as citizens of the global community, it is our individual and collective responsibility to uphold a sustainable way of life for the coming generations.

But how do we do that?

Cloudy Days and Silver Linings

Despite the harrowing headlines and the rapidly deteriorating nature of the situation, I am happy to say that there is a silver lining to be found. With so many fighting to make even the smallest change in their everyday lives, the positives have finally begun to compound, and there is real hope for the revival of our species and for our planet.

While it is unlikely that we will achieve the ambitious goals set by the Paris Agreement, this will not be the end of our lives.

Granted, as temperatures rise, we will see troubling changes in our atmosphere—things will get hotter, colder, wetter and drier, depending on where you are in the world. Extreme natural events like heatwaves, wildfires, cyclones and typhoons will become more commonplace and will burden societies and natural ecosystems even more than before.

As these structures collapse under the weight of the surmounting pressure, underdeveloped nations will face famine and severe droughts, leading to widespread damage and the loss of our ability to sustain a population of over eight billion people worldwide.

Around the world, many species and people will be lost to these near-apocalyptic events. But does this mean the end times are here?

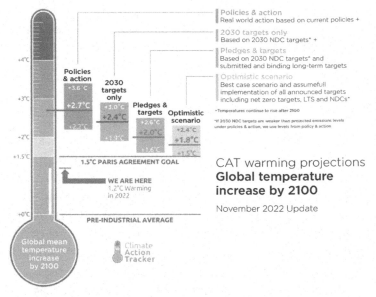

Figure 1-2: Climate Action Tracker warming projections about global temperature increase by 2100. Source: Climate Action Tracker[6]

I am an optimist, and I have seen the world change many times over without crumbling under the pressure of what is to come.

Years ago, when Al Gore's *An Inconvenient Truth* shone a light on the dangers of rapid climate change, many dismissed the issue as another apocalyptic prophecy that would never come true. As time went on, the narrative changed. People accepted climate change was a threat, but would not accept responsibility for their actions. Now, we have entered the final phase of our cycle of denial—acceptance. Climate change is here, we are responsible and only we can save ourselves.

While things may not stay the same, humanity has gotten as far as it has solely because of its ability to evolve. While we will indeed have to brave losses and change our ways, we are likely to endure the coming climate crisis.

The Lost(?) Decade of Climate Action

As you are probably already aware, the last decade saw a raging battle for the conservation of our planet. With one failure after another besieging media headlines and stagnating policy growth towards sustainable lifestyles, and with the rampant and relentless misinformation campaigns spearheaded by the capitalist overlords that benefit from the destruction of our planet, it is not hard to imagine why so many struggle to see a future abounding with opportunity.

However, change does not always announce its arrival with a megaphone. In most cases, it is an invisible force, one that you don't see coming until it is at your doorstep, knocking at the door with a friendly smile and a warm demeanour.

A great example of this is the correlation between a nation's emissions and its ability to grow its economy. Once, fossil fuels were offered to underdeveloped nations like India and China at a subsidy, to allow them access to an affordable route to progress. However, the emissions associated with this growth were widely condemned, with developed nations demanding greater action on the part of these developing nations, while doing little to limit their own emissions.

This created an irrefutable narrative—to develop, you must emit more.

However, the following years painted a more promising picture.

Developing nations like China and India found a way to neutralize their emissions, leveraging technologies that had never existed when developed nations like the UK and the US were in their younger days.

With technologies for renewable energy rapidly evolving and becoming more cost-efficient, today, it is actually *more* expensive for countries to stick to coal and fossil fuels for their energy.

Globally, the generation of renewable energies has increased, and technologies that once seemed too expensive or too complex to integrate into our everyday lives have reached maturation, becoming more accessible and affordable every day.

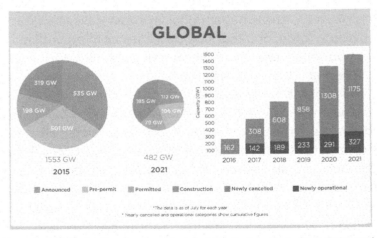

Figure 1-3: Reduction in size of global coal project pipeline (left) and year-on-year tracking of projects that were cancelled (right).[7] *Source: Global Energy Monitor*

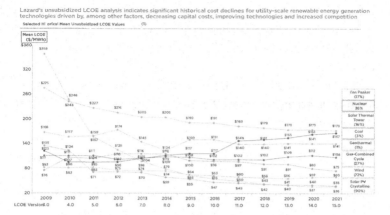

Figure 1-4: Levelized cost of energy consumption--historical utility-scale generation comparison.[8] Source: Lazard

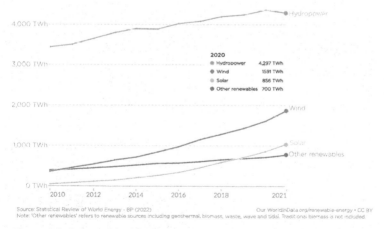

Figure 1-5: Renewable energy generation[9]. Source: Statistical Review of World Energy

In this new world, coal, which once held a monopoly over the energy market, is now not competitive enough to beat its green counterparts.

Green innovation is on the rise globally, with tech companies racing against each other to get to the finish line. Unlike the age of coal, in this race, we are all winners.

This means that cutting down on emissions is no longer a straight road to recession, but implies that developing countries are skyrocketing into the future using these new technologies to skip past the errors of those who came before, making sustainable growth and economic development a reality we are already witnessing around the globe.

Amplifying this invisible, positive force for change, other global movements, like the shift to electronic vehicles over internal combustion engines, as well

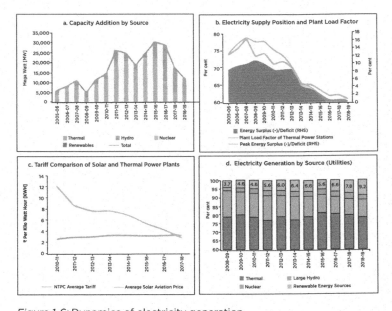

Figure 1-6: Dynamics of electricity generation

as a plethora of innovations towards mitigating the damage of oil spills, carbon capture devices and hundreds more, are causing a ripple effect, motivating individuals to live more sustainably and demand more from their politicians and legislators.

Globally, we have reached a point where going green is not just a good-to-have, but is also a good economic move, with unsustainable and damaging industrial behaviours being increasingly and widely condemned by a growing population of conscious consumers every day.

While this trend does indeed paint a promising picture, we must be conscious that just because we're on the right track, it doesn't mean we are in the clear.

While technologies are rapidly evolving, they are not moving nearly fast enough. As a species, we are still doing too much damage with too little to stop it, and scientific research and technological development alone cannot clean up our mess.

But if we were able to achieve so much with so little support from our governments and industries, imagine what we could achieve if we worked together.

According to Dia Mirza, United Nations Environment Programme goodwill ambassador and actress:

> '"All of our solutions have always been in nature" is a message I have heard numerous times on different forums, but I see a disconnect between mass consciousness and the very real threat of rapid climate change, which is the most pressing issue of our times. I would urge everyone to remember that

together we can reimagine, recreate and restore our beautiful earth, on which our health and survival depend. We must join movements that are invested in the protection of the planet and demand laws that will ensure the change we need. When political, business, and environmental leaders work in synergy with people from all walks of life to protect the earth, a better world will come into being.'

In Defiance of Doomerism

Defined as an attitude of extreme pessimism towards the issue of climate change, and the dwindling possibility of ecological rehabilitation, Doomerism is the single greatest enemy in the fight for sustainability.

With so much media coverage twisting the words of well-meaning scientists cautioning us about dire consequences for our collective future, the fossil fuel industry as well as all those who benefit from unsustainable consumption and production have weaponized hopelessness to their benefit.

How, you ask?

From BP to Coca-Cola, the largest capitalistic polluters of our natural ecosystems have crafted a clever narrative—that their unsustainable actions are in service of their consumers and so, it is the consumers themselves who must make more sustainable decisions. From large-scale greenwashing campaigns urging us to lower our individual carbon footprints, to the call-to-arms for consumers to go plastic-free even as plastic continued to reign over the packaging industry, the onus was on us to change the way we live.

However, while we can indeed limit our emissions and try and eliminate plastic from our everyday lives, we must also realize that these billion-dollar industries that profit from unsustainability will never be rectified if not brought to the court of public opinion.

No matter how hard we try, we can only do so much. If our governments truly wanted a change, change would be here already. If the policies were going to make a meaningful difference, they would have already.

In a world where we are held responsible for the actions of others, and where our efforts to reconcile with the natural world fail time after time, due to forces outside our control, we cannot help but feel hopeless at the prospect of a bleak future.

But this hopelessness breeds fear, sadness and uncertainty. As the uncertainty grows, more and more of us have resigned to live out our days however we can and brace for the inevitable collapse of humanity.

However, this apathetic, indifferent attitude towards our natural world is exactly what the fossil fuel industry wants. The longer change is delayed, the longer they can protect their economic standing and delay making costly changes to their institutions.

What we need to realize is that these institutions were not built by the giants themselves, but by us. If we can build empires by purchasing one bottle of fizzy soda at a time, we can raze these empires to the ground, one conscious choice at a time. The powers that be may have shifted the blame onto us, but they were right about one thing—the consumer holds all the power, and we are a force to be reckoned with.

The best way to truly fight climate change is to fight hopelessness. We must know that we are not mere spectators in the decline of our global civilization, but have the power to affect change at every level of industry, politics and society.

But how do we do that?

Climate Optimism

The answer is simple—we raise our voices. We become optimists who fight not for a future that is already lost, but one with uncountable opportunities to completely transform our relationship with the natural world.

But what is climate optimism in a world where doomerism rules supreme? When most people think of optimism, they imagine that it means a state of unwavering belief that is unrealistically positive.

However, to me, optimism is the belief system that prioritizes the fairer side of the coin. There will always be a good and a bad side to life, but what we choose to focus on is our choice, and ours alone.

Climate optimism doesn't deny the reality that we face every day, but it is a way of thinking that takes in the good and the bad and allows us to understand how we can change so that there is more good to focus on. This is because optimists do not retreat to safety at the slightest hint of danger—they can take the negatives with the positives, and focus on building constructively towards a better and brighter reality.

I do not ask you to pretend that the world is all roses and sunshine, but rather would like you to remember the beauty of sunshine, and the sweet smell of roses, and that protecting their beauty is not a matter of drastic action, but slow and sustained effort towards preserving all that is beautiful in the world.

Our voices and choices are our greatest weapons in the battle against climate change, but even greater still is the virtue of hope.

We are at a crucial moment in our fight for hope and for our planet. While the silent momentum is promising, it must continue with even greater inertia, building up the pressure until the levee breaks and we can stand with our planet, not in defiance of it.

This book may not be the answer or the perfect guide to show us how to overcome the challenges of our time—but it is a starting point. It is the 'You Are Here' symbol on the map to progress, and it is up to us to move in the right direction.

The landscape of sustainability and innovation to combat climate change is rapidly evolving, with new answers and new questions being raised every day. Every solution presented in this book will have its downsides; there is no perfect solution to the crisis we face. But this is okay—after all, the goal is mitigation, not eradication. Our lives, our planet and the universe as a whole are subject to yin and yang, with the forces of good and evil constantly pulling against each other, keeping us in a state of perpetual motion.

We might not have all the right answers, but we do have hope. Every single choice we make has the power to cause a ripple effect that extends infinitely outwards into the universe, and sowing even the smallest seed promises to bear fruit in the times to come.

Time and time again, we have proven that we can beat the odds and overcome even the direst circumstances. Well, my friends, that time is here again, and we must unite to tip the scales in our favour.

It is in our nature to focus on the problem, but the solution exists and is a far more rewarding notion to spend our time on. If there is one thought I would like to leave you with before we begin this book, it is this— **we can solve this problem**.

Being optimistic, and seeking hope in the darkness, is not just a practice that will help us chart a better course for the future, but is also a solution for the mental health crisis that plagues us today. It allows us to re-centre, to focus on what is working and the progress that has got us to where we are today. While the world is full of things that promise to tear us down and wither away our defences, we must stand united and support each other towards the future that we want to see.

When Mahatma Gandhi wanted to free our people from the colonial regime, he did not wait for someone to tell him the time was right. He did not look to the authorities or the politicians for help. He looked to his people and asked them to march with him in the light of truth, and in the promise of hope.

Today, I ask you to do the same.

I am an optimist, and I believe the future is ripe with possibility. However, to avail of that future, you and I must fight tooth and nail against the hopelessness of our times, and be a guiding light for those who will come after us. The onus is on us to build a new world, but in this new world, there is no space for negativity.

Let me show you how, through the powers of optimism, belief, collaboration and a little bit of science, we can change the world, one person at a time.

How to Read This Book

Before we begin, here's a note from the author:

This book is not like other books on climate change or global warming. It is not a novel or a guide, but something in between. This book is a starting point for anyone looking to be the change, but I cannot in good conscience say that it contains every answer. This is because climate research is evolving rapidly and innovation is frequent. What we know today might not be the answer that will lead us into tomorrow.

Keeping this in mind, every chapter in this book tackles a real-world issue from a global perspective and is structured into four major sections:

The Problems: The overarching, underlying issues that we may not at first realize.

The Solutions: The actions we may take to combat the unknown, and the unsustainable, today.

The Profile: Spotlighting grassroots changemakers, who are living in harmony with the natural world and whom we can learn from.

Further Reading: Since climate action is an ongoing process of research and innovation, every chapter contains its own set of resources for those looking to dive deeper, learn more and make actionable changes.

I encourage every one of you to join me on this journey, and along the way, try and find new paths to walk, new solutions to innovate and discover hope, so that others may one day follow suit.

2

Food and Beverages

Greenhouse Gas (GHG) Impact: 31 per cent[1]

This industry impacts the following Sustainable Development Goals (SDGs):

- SDG 2: Zero Hunger
- SDG 3: Good Health and Well-Being
- SDG 6: Clean Water and Sanitization
- SDG 10: Reduced Inequalities
- SDG 12: Responsible Consumption and Production
- SDG 13: Climate Action
- SDG 15: Life on Land

ACTIVITY

Before we begin this chapter, let's try to respond to the following questions as honestly and thoroughly as possible:

1. How many of the following do you consume as a part of your diet?
   - ☐ Milk
   - ☐ Cheese
   - ☐ Avocado
   - ☐ Aerated beverages
   - ☐ Coffee
   - ☐ Tea
   - ☐ Chocolate
   - ☐ Nuts
   - ☐ Lentils
   - ☐ Poultry
   - ☐ Red meat
   - ☐ Fish
   - ☐ Eggs
   - ☐ Flour

2. Which of the following do you consume more than the other?
   - ☐ Avocado/mushrooms/neither/both
   - ☐ Eggs/fish/neither/both
   - ☐ Potato chips/banana chips/neither/both
   - ☐ Coffee/tea/neither/both
   - ☐ Frozen food/fresh food/both

3. In a week, how many times do you order in?
   ☐ Less than one to two times a week
   ☐ One to two times a week
   ☐ Three to four times a week
   ☐ Six or seven times a week

4. Which of the following do you do with your food waste?
   ☐ Composting
   ☐ Recycling
   ☐ Donation

5. Where do you buy your food from?
   ☐ Supermarket
   ☐ Local marketplace
   ☐ Online
   ☐ Other (please specify)

The history of human evolution is also the chronicle of the food we consume.

As cave-dwellers, we were hunter-gatherers, eating only what we could kill and what we could pluck from trees. Later, as we took to agriculture, our consumption habits evolved.

Today, food holds far greater value than mere sustenance. It's not only the expression of our mood and sentiment but also an important component of our festivals. From sugary-sweet icing on birthday cakes to aromatic biryanis on Eid, and from the *bhaang*-infused *thandai* of Holi to the communion wine and bread of the Eucharist, the food we eat is so inexorably intertwined with tradition and custom that it serves as a part of our cultural identities. It comes as no surprise, then, that an assault on our consumption patterns is viewed as a criticism of our way of life.

In the contemporary world, the global food and beverage manufacturing industry employs an estimated twenty-two million workers, with the global food and grocery retail market size being valued at $11,324.4 billion in 2021 alone.[2]

From avocado toast to juice cleanses, humanity has come a long way from its Palaeolithic way of life. Today, we are privileged enough to *choose* what we consume.

However, in this quest for flavour, health and status, humans have kept themselves at the centre of the endeavour, neglecting the impact of their consumption on the planet and all those who inhabit it.

While we currently produce more than enough food for the global population, the United Nations states that 8.9 per cent of the world's population, i.e. 690 million people remain hungry.[3] This figure is even more unsettling when one takes into account that, 30 to 40 per cent of the food produced in the world is lost before it even reaches the commercial market, for a variety of reasons ranging from improper storage and processing to transportation and more.[4] According to the Food Waste Index Report 2021 by UNEP, India alone wastes 50 kg per capita per year.[5]

When I was a child in the early 1990s, my family, like many people in this nation, faced rainy days as well as easier ones. I still remember the harder days, when we were unable to afford vegetables to make a decent meal. As a nomadic traveller, I have gone days without the mere sight of food.

I know from personal experience what true hunger feels like. I know the pain in your belly and the weight in your chest that comes with not having the means to feed yourself.

That is why these jarring figures got me thinking—if we had to draw up a meal plan that protected the health of our planet as well as all those who inhabit it, what would we eat?

What I found over the course of my research changed the way I look at food forever.

For most of us, consumption habits and preferences can be traced back to our early childhood, inculcated within the familial systems we occupy. In other words, we eat what our families, friends and neighbours eat.

I grew up in a vegetarian family. Like most Indian mothers, my mother wanted me to grow up to be a

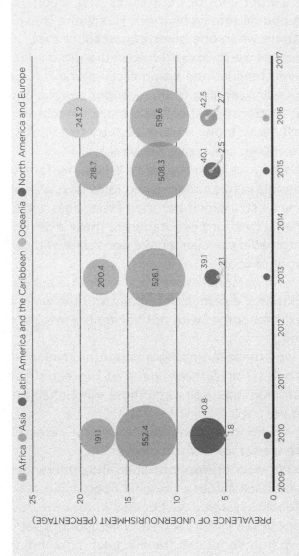

Figure 2-1: The prevalence of undernourishment is highest in Africa; the absolute number of undernourished people is largest in Asia.[6] Source: Food and Agriculture Organization

NOTE: Comparison of prevalence and number of undernourished people by region. The size of the circles represents the number of people in millions, as labeled. Figures for 2016 are projected values.

strong, healthy boy and would goad me to drink a glass of milk every day. While I didn't like the milk, I must admit that I loved sweet *dahi* (yoghurt) and cherished the bowl I would devour with my meals. My chapatis were served slathered with ghee, and whenever we'd go out for dinner, paneer sabzi was a must-have, followed by dessert.

Can you see anything wrong with this diet? **I couldn't**.

As far as I understood it, my diet did not kill any animals for their meat, and we paid for our indulgences with hard-earned money—what could be wrong with that? In my eyes, I was doing the right thing and living in harmony with society and nature.

But what if I were to tell you that both vegetarian and non-vegetarian diets are unsustainable in the long-term, and are responsible for the destruction of not only our environment but also of our health?

The Uncomfortable Realities of 'Comfort Food'

The phrase 'comfort food' is a contradiction that succinctly captures the state of consumption in the twenty-first century. To illustrate this point, let's take the example of the comfort food that has existed for over four millennia, and is consumed by a large majority of the world's population even today—chocolate.

From Roald Dahl's *Charlie and the Chocolate Factory*[7] to hot cacao in the winter months, chocolate has been synonymous with comfort, indulgence and the simple joys of childhood. However, according to a report[8] published in 2018, the global chocolate industry contributes around 2.1 million tonnes of greenhouse gases (GHGs) to the atmosphere every year.[9]

The cultivation of cacao, the main ingredient of chocolate, is also responsible for hectares of deforested land every year, with forests lost due to cacao production estimated to be between two and three million hectares between 1988 and 2008.[10] These rainforests are home to several endangered species, including orangutans, tigers, elephants and rhinos. Between 2001 and 2014, the Mighty Earth Report recorded 2,91,254 acres of protected forests cleared in Cote d'Ivoire alone.[11]

Chocolate production is also highly water-intensive, with a single kilogram of chocolate utilizing about 10,000 litres of water.[12]

While cacao isn't produced in India, Indians account for a tenth of global confectionery consumption,[13] with consumption per capita at 100g to 200g per person.[14] This means that a majority of our chocolate must travel across oceans and land to reach us, adding to its carbon footprint with every kilometre travelled, with an average bar of chocolate carrying a carbon footprint of approximately 3.45 gm of carbon dioxide per gram of chocolate.[15]

This is not to mention that the chocolate industry is highly reliant upon the dairy industry, which, in turn, is responsible for 4 per cent of total anthropogenic GHG emissions.[16]

If this is the first time you have had the chance to think about what goes into that sugary-sweet bite of chocolate, you are probably shocked. I know I was.

The typical reaction to this information is guilt—people blame themselves for what they did not know.

But I'm going to tell you something really important: *it's not your fault you didn't know.*

Big corporations allocate massive advertising and public relations budgets to ensure that we, as consumers, are kept in the dark. Lakhs of rupees each year are invested in creating the fantasy of 'comfort' in comfort foods, while the reality of environmental degradation, which can negatively impact sales, is hushed.

When reading this chapter, it is important to remember that **you are not ignorant**, these facts are **deliberately buried**, labelled 'industry secrets' by those who prioritize profits over natural welfare and sustainability.

I know what you're thinking:

If this information is such a secret, how do I know about it?

*

My relationship with sustainability was completely transformed in 2017 when I was living in a small village in the Spiti Valley, a remote trans-Himalayan region in the northern tip of India, which shares a border with China. A nomad by nature, I found myself drawn to the serenity and remoteness of mountains, far away from the hustle and bustle of metropolises. Here, I could be closer to nature and share meaningful exchanges with fellow travellers through the region.

At the time, I was a self-declared student of sustainability, learning as much as I could about the complexities of climate change and how I, as an individual, was contributing to it. It was so that I happened to come across a ground-breaking documentary co-directed by my now dear friend, Keegan Kuhn. An incomparable narrative chronicling the harsh realities of animal agriculture, *Cowspiracy: The Sustainability Secret*[17] is a documentary that changed the way I looked at food forever.

Over an hour and thirty minutes, I learned things that I would never have even considered possible until this point. Suddenly, an intricate web of unsustainable practices unfurled before me—a web that I found myself caught in, along with so many others.

For instance, the dairy industry emerged as one of the leading sources of GHG emissions due to foraging, which requires hectares upon hectares of deforested land. Methane, which accounts for approximately 25 per cent of global warming[18], and whose CO_2 equivalent is approximately 84x[19], is also produced in vast quantities on dairy farms through the cattle's natural bodily functions, fermentation practices and manure storage.

I learnt about palm oil, a standard ingredient in a wide range of food and beauty products, which accounts for the loss of 300 football fields worth of rainforests every single hour.[20] Even my seemingly innocent morning coffee became almost sinister in its implications.

The more I researched, the further the web stretched.

This was just the tip of the iceberg, but it is also the unfiltered reality of our times.

Director Keegan Kuhn, who has directed films such as *Cowspiracy* and *What the Health*, says:

> 'No other industry has a further reaching impact as animal agriculture. Raising animals for their flesh, milk, eggs and skins, is the leading driver of deforestation, water consumption, water pollution, ocean dead zones, ocean plastic (fishing nets), topsoil erosion, species extinction, desertification, habitat destruction and a primary contributor to climate change. Virtually anything you can care about in the world, animal consumption plays a major role in its destruction. Never in the history of the planet has there ever been 8+ billion megafaunas of a single species existing at once. We have a right to be here, but not everywhere at once. We need to allow the wild ones to have their own space. I have dedicated most of my life to promoting environmental knowledge. I think people need all the information to make informed decisions.'

It was almost too much for me to grasp, and accept, over the course of an hour and thirty minutes. Over the next ten days, I watched the documentary seven times. On the eleventh day, I pledged to adopt a plant-based lifestyle, giving up the vegetarian diet I had inherited from my childhood.

Going plant-based is not easy, but I found that it was harder for those around me to accept this step I had taken. My friends teased me, saying I had abandoned flavour in exchange for dry wisps of half-baked nutrition. My friends and family cautioned me against it, claiming it was an impulsive decision that would impact my health long-term, leaving me weak and of fragile disposition.

However, five years later, my relationship with food, and with consumption overall, has transformed into one that is healthier, and more sustainable for the planet. I run, cycle, trek, kayak and push my body to its limits every day, and I have never once felt the need to return to my old diet.

Since then, I have transformed my lifestyle completely to live in the most sustainable way possible. I do not consume products that cause harm to any living creature anywhere along the supply chain, and I only shop from brands that prioritize ethical sourcing, manufacturing, production and sales—I live closer to nature and am grateful for the many gifts it has given me.

Over the last five years, I have grown from a student of sustainability to an advocate of it, and I owe much of my drive and conviction to my plant-based lifestyle.

As an advocate, I believe it is my duty to share what I have learned, so that my words and actions may inspire someone in the way Keegan was able to inspire me.

Let me tell you a little bit about my journey, and how, over the years, I have managed to keep the integrity of nature central to my plans.

The Problem with Our Pantries

Try this: walk into your kitchen and open your pantry. What do you see there?

Land Use Change | **Farm** | **Animal Feed** | **Processing** | **Transport** | **Retail** | **Packaging**

above ground changes in biomass from deforestation, and below ground changes in soil carbon | Methane emissions from cows, methane from rice, emissions from fertilizers, manure, and farm machinery | On farm emissions from crop production and its processing into feed for livestock | Emissions from energy use in the process of converting raw agricultural products into final food items | Emissions from energy use in the transport of food items in country and internationally | Emissions from energy use in refrigeration and other retail processess | Emissions from the production of packaging materials, material transport and end-life disposal

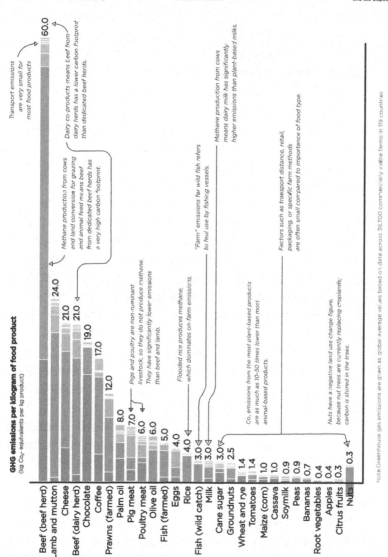

Figure 2-2: Food/Greenhouse gas across the supply chain.[21] Source: Our World in Data

If your diet consists of meat and dairy products, you will likely find some milk, cheese, yoghurt, eggs, chocolate and an assortment of meat products best suited to your palate. Chances are, you are also likely to find non-dairy and meat products, such as rice, pulses, fruits, vegetables, olive oil and bread.

Now, take a look at the following graph.

This is the reality we must now confront.

Sometimes, even the choices we think are the right ones can be harmful to us in the long term.

But can a plant-based diet put a stop to this? *The simple answer is* **yes**!

Did you know that, with the current state of affairs, the global food and beverage production industry produces pollution in the form of GHGs to such an extent that if all the nations in the world agreed to cut all non-food related emissions to zero, we would still be unable to limit the rise in global temperatures to 1.5°C—the target agreed upon in the landmark Paris agreement?[22]

However, if we as consumers converted to a more plant-based diet on average, while simultaneously cutting emissions from other sectors, we would have a 50 per cent chance of success.[23]

Moreover, if we made more conscious decisions within our food systems overall, such as limiting food waste, our chances of meeting that target would increase to 67 per cent.[24]

In much the same way as one unsustainable choice sets off a chain of destruction that affects millions across the globe, so too does one step in the right direction cause a domino effect that can change the lives of many for the better.

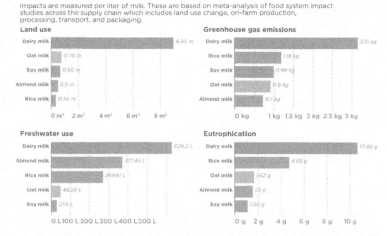

Impacts are measured per liter of milk. These are based on meta-analysis of food system impact studies across the supply chain which includes land use change, on-farm production, processing, transport, and packaging.

Land use

Dairy milk	8.95 m
Oat milk	0.76 m
Soy milk	0.66 m
Almond milk	0.5 m
Rice milk	0.34 m

0 m² 2 m² 4 m² 6 m² 8 m²

Greenhouse gas emissions

Dairy milk	3.15 kg
Rice milk	1.18 kg
Soy milk	0.98 kg
Oat milk	0.9 kg
Almond milk	0.7 kg

0 kg 1 kg 1.5 kg 2 kg 2.5 kg 3 kg

Freshwater use

Dairy milk	628.2 L
Almond milk	371.46 L
Rice milk	269.81 L
Oat milk	48.24 L
Soy milk	27.8 L

0 L 100 L 200 L 300 L 400 L 500 L

Eutrophication

Dairy milk	10.65 g
Rice milk	4.69 g
Oat milk	1.62 g
Almond milk	1.5 g
Soy milk	1.06 g

0 g 2 g 4 g 6 g 8 g 10 g

Figure 2-3: Environmental footprints of dairy and plant-based milks.[25]
Source: Our World in Data

For instance, by simply switching to soy, almond or oat milk instead of cow or goat dairy products, you would cause a chain reaction that positively affects not just the animals themselves, but also our planet's water supply, air quality and even our natural biological systems.

How, you ask?

Bred in commercial numbers for their milk and meat, the animal husbandry industry utilizes more than 20 per cent of the world's fresh water.[26] In addition to the cruelty meted out to these innocent animals— including hormone treatments and despicable living conditions—the meat and dairy industry is responsible for rapid forage-related deforestation across the globe.[27]

Kiran Ahuja, manager of vegan products, PETA India, says:

> 'Humans are eating more animals today than we ever have in history. Worldwide meat consumption more than doubled from 1950 to 2009. Today, we kill eighty billion land animals (about 8x the entire human population) and up to trillions of fish for food. Each one of these numbers represents a thinking, feeling individual who did not want to die.'

The Red Sindhi cow is one of the most distinctive breeds in India, found in the regions of Punjab, Haryana, Karnataka, Tamil Nadu, Kerala and Orissa. Under fair management conditions, the Red Sindhi averages over 1700 kg of milk per lactation after suckling their calves. However, under optimum conditions, milk yields have been recorded to cross over 3400 kg per lactation.[28] In 2018 alone, India consumed approximately 66.8 million metric tonnes of milk.[29]

The growth of the Indian population has been on a steady rise, and with the consumption of milk being an important part of rural and urban culture, the demand for dairy products is on the rise as well. To meet this growing demand, the dairy industry uses a variety of invasive and cruel methods to boost the animal's natural milk production cycles, including robbing mothers of their young, artificial insemination, delivering growth hormones to boost milk production, injecting drugs such as oxytocin and much more.

The cruelty doesn't stop there. To the industry, these animals are but a means to an end. Milk production

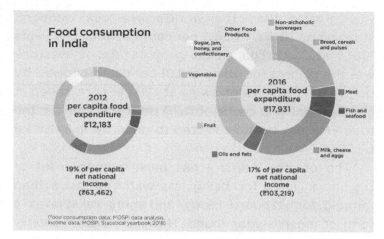

Figure 2-4: Food consumption in India.[30] *Source: Government of India*

is the goal, and anything that doesn't serve this goal must be left behind. This is the justification used when slaughtering male calves, abandoning sterile animals and selling livestock off to slaughterhouses and tanneries when they are too old or too tired to produce milk.

India is home to the world's largest population of livestock. Every year, India produces approximately 5.3 million metric tonnes of meat and 75 billion eggs.[31] In terms of global meat production, India is the largest producer of buffalo meat as well as the second largest producer of goat meat. While India has banned the sale and consumption of beef in many of its major states, it is the third-largest exporter of beef in the world.[32] It has been estimated that over 200 million tonnes of CO_2 equivalents are released into the atmosphere by Indian livestock each year.[33]

In terms of methane, the percentage increase in enteric methane emission (EME) by Indian livestock was greater than the world's livestock quotient, standing at 70.6% vs 54.3 per cent between the years 1961 and 2010. The annual growth rate of livestock was the highest for goats at 1.91 per cent, followed by buffalo at 1.55 per cent, swine at 1.28 per cent, sheep at 1.25 per cent and cattle at 0.70 per cent. In India, the total EME has been projected to grow by 18.8×10^9 kg by 2050.[34]

The poultry industry has been gaining a lot of traction in India for being less water-intensive than farming for standard crops, and more reliable as a form of income year-round. However, Indian poultry farms also have a reputation for low health and safety standards and are a major source of public health concerns.

According to the 20th Livestock Census, India is home to 851.8 million poultry birds, about 30 per cent, i.e., 250 million of which are categorized as 'backyard poultry', being raised by individuals and marginal farmers.[35]

Investigations conducted into the conditions at Indian poultry farms revealed a multitude of cruel and unhygienic practices, including, but not limited to, hens covered in painful sores and housed in unsanitary conditions marked by faeces, spiderwebs, mites and lice.[36]

The feed being provided to the hens was found to contain a variety of heinous and concerning ingredients, including chips of marble, cardboard, grains laced with heavy doses of antibiotics and pesticides, fish meal, and perhaps most troublingly, remains of dead chicken. The nutritional quality of the poultry feed is in direct correlation to that of the eggs they produce, leading to widespread contamination.

In fact, despite being one of the three largest producers in the world, with 47 billion eggs produced annually, Indian eggs fall short of international standards and cannot be exported due to chemicals found both in and outside the eggshells.[37]

In a recent study conducted around Hyderabad, eggs were sampled both from retail outlets as well as directly from poultry farms. In nearly all samples collected from retail outlets, researchers detected traces of the highly infectious and lethal Salmonella bacteria, known to cause such medical health emergencies as typhoid fever, food poisoning, gastroenteritis and enteric fever.[38]

According to the Water Footprint Network, it takes 1,020 litres of water to produce 1 litre of cows' milk; 3,265 litres of water to produce 1 kilogram of eggs and 15,415 litres of water to produce 1 kilogram of beef. In stark contrast to these figures, it takes just 322 litres of water to produce 1 kilogram of vegetables.[39]

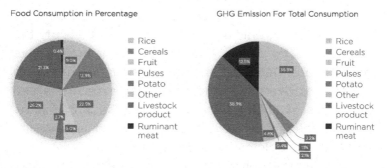

Figure 2-5: Food consumption in percentage and greenhouse gases for total consumption.[40] Source: Science Direct

The rearing of livestock in such vast numbers is also a public health hazard, as demonstrated most recently by the COVID-19 pandemic of 2020. According to a report by UNEP and ILRI, about 60 per cent of known infectious diseases in humans and 75 per cent of all emerging infectious diseases are zoonotic in nature and result from improper housing of livestock and the storage and sale of meat products in wet markets.[41]

But what about fish?

Seafood is often promoted as a healthier option than poultry or red meat, both for our health as well as for the planet.

In coastal areas, such as India's Maharashtra, Tamil Nadu and Kerala, seafood is a staple part of the area's diet and is consumed with most meals. For pescatarians and those looking to reduce the amount of meat in their regular intake, fish serves as a viable source of protein.

However, seafood consumption has damning effects on both the sanctity of our oceans as well as on the natural systems that rely upon it.

Seaspiracy[42], the 2021 documentary by director Ali Tabrizi, chronicles the harsh realities of the fishing industry and is a must-watch for prospective pescatarians. Dumping of sewage and industrial waste into oceans, rivers and other water bodies has increased the levels of chemicals and microplastics in marine animals, leading to devastating losses of aquatic life and health problems for those who consume catch.

In India, the problem is so severe that in 2021, several foreign nations stopped importing seafood sourced from Indian waters due to high levels of lead and cadmium, according to an investigative study conducted across 241 fish farms in ten Indian states.[43] [44]

As per the study, Tamil Nadu fish farms recorded the worst quality of water, while Andhra Pradesh, West Bengal and Pondicherry farms exhibited high levels of public health hazards. Of all the farms studied, Tamil Nadu, Bihar and Orissa were found to be the most harmful to the environment.

As such, consuming commercial catch is likely to fill your body up with heavy metal pollutants and chemicals such as methylmercury, PCBs, dioxins and more. Not to mention, microplastics and fibres from waste discarded in the sea years ago could land up in your stomach.[45] The prevalence of microplastics in our water, seafood and more is so high that microplastic particles were found in the placentas of unborn babies as recently as 2020.[46]

Industrial fishing also leads to environmental degradation by depleting the complex ecosystem of the oceans, which acts as a vital climate regulator. The problem isn't just with how much we are fishing, but how we are fishing. Bottom trawler nets with weighted, steel fittings drag across coral forests and reefs, effectively bulldozing everything in the way. These coral forests are slow-growing in nature, and take years and years to grow to their optimal state to support other life. When these ancient ecosystems are destroyed, reviving them is no simple task.

However, the existence of these vulnerable natural ecosystems and the biodiversity they support are vital to fish stocks globally, as well as to our climate.

An ocean brimming with life and diversity means higher rates of carbon sequestration—a vital natural process that locks carbon emissions. The ocean is the largest reservoir of carbon on our planet, with an estimated 38,000 gigatonnes of carbon already stored.[47]

This carbon is stored by the living creatures within the ocean, which capture and store the emissions that would otherwise contribute to global climate change. The oceans are thought to be the only net sink of human CO_2 emissions over the last 200 years.

Giving up seafood, then, will not only help improve your health but also reduce the environmental impact associated with commercial fishing. Overfishing is depleting the marine population the world over, and adversely affecting indigenous coastal communities who depend on fishing for their livelihood. Fish feed manufactured for the commercial fishing industry, in turn, causes mass deforestation.[48]

But let's take a step back and return to our kitchen pantry. What else do you think could negatively impact the sanctity of our environment?

Do the names of famous soft drink brands pop up in your mind?

While our parents have long warned us against the unhealthy lifestyle choice of a can of soft drink with dinner, aerated beverages like soda pops and fizzy drinks put a greater strain on our natural resources than we may imagine.

In 2021, a popular soft drink company recorded a water use ratio of 1.94 litres of water per litre of product produced.[49] According to a report by Good

To Know, every 100 ml bottle of Coca-Cola contains approximately 10.6 grams of sugar.[50] This begins a vicious cycle—the more you drink, the thirstier you get, and the more product the corporation sells. As Statistic.com reports, in 2021, the company sold **31.3 billion** unit cases compared to 29 billion unit cases in 2020.[51]

So the next time you are looking for a refreshing drink to cool off in the summer, remember that you are better off drinking a litre of water instead.

If all this talk about capitalism and the food industry is making your head spin, why not relax with a cup of piping hot tea, or perhaps coffee?

I take no joy in proclaiming that coffee and tea are not innocent, either.

If you ever visit a plantation, you would soon realize that they function as monocultures, meaning that only a single crop is grown over vast expanses of land. These water-intensive crops have a dark history of human exploitation that can be traced back to colonial times, and harm our planet's natural integrity even today.[52]

If, as a coffee drinker, Kopi luwak—one of the world's most expensive types of coffee—has been on your bucket list, you might want to reconsider.

The coveted coffee is produced from beans digested and excreted by the Asian palm civet who are held captive for profit and tortured for our benefit.[53] Even some of the world's largest food and beverage companies conduct—and pay others to conduct—painful and often fatal tests on animals to give you that perfect cup of tea.[54]

Now, is that really how you want to start your morning?

According to Manoj Kumar, co-founder of Araku Coffee:

> 'You cannot have a sustainable food system if the producers are always making losses. So to me, the core of sustainability comes from ensuring the producers make sustained profits. If we don't build an economic model on that, then trying to sustain the planet is virtually impossible, with no incentive for the producer to protect our planet. The Araku model is about making farmers to be entrepreneurs, to ensure for them consistent profits and agency.'

What about the packets of instant noodles that line your shelves, or the frozen foods and ice creams that clutter your freezer?

Frozen foods have gained a lot of traction in India over the last decade, going from frozen peas and French fries to a vast array of ready-to-make products ranging from the exotic to the everyday.[55] This is largely due to the invention of speed freezing, a technological process that allows for rapid freezing at mass scales.

Take peas, for example—every pea must be individually frozen, which is a time and resource-intensive process. With the advent of the age of speed freeze foods, however, a pea can be harvested, transported, machine washed, blanched and frozen in a span of just two hours.[56]

FROZEN PRODUCTS- RETAIL	2013		2020	
	Rs. Cr.	CAGR	Rs. Cr.	CAGR
Sea Food (Processed & Raw)	105	10%	205	14%
Meat & Meat Substitutes	50	10%	105	16%
Poultry Meat	760	18%	2100	25%
Non-Veg Snacks/Meals	65	22%	200	30%
Veg Snacks/Meals	60	21%	150	21%
Vegetables	260	10%	475	12%
Potato Products (incl FF)	60	20%	185	30%
Frozen Desserts	40	7%	80	14%
TOTAL Frozen Packaged Food	1400	15%	3500	21%

Table 2-1: Frozen food and compound annual growth rate (CAGR).[57]
Source: Renub Research

This is important because speed is of the essence in the frozen food industry: as soon as produce is taken from the ground, it's a race against time to preserve its nutritional quality, with every step of the process robbing the produce of some of its value. By the end of the cycle, however, much of that nutrition is lost, no matter how quickly it is executed.

As consumers, we are duped into opting for variety over nutrition, chasing exotic and flavourful frozen offerings over fresh foods that lack the convenience speed freezing provides.

Okay, so frozen foods are bad for our health, but surely, they could not be as environmentally destructive as meat products, right?

Try this:

Flip your packets of instant noodles, frozen and other processed foods over, and analyse the ingredient section. Do you see the common thread that binds them all together?

Palm oil is a tasteless, low-cost resource and an incredibly efficient crop that yields a much higher harvest per area of land than coconut or soybean. This is part of the reason it is a common ingredient in most processed foods, as well as beauty products, like shampoos and soaps.

But palm oil is made from palm—doesn't that mean it is plant-based?

Technically, yes. Palm oil, when consumed in its raw form, is plant-based. However, the increasing demand for palm oil is one of the leading causes of environmental destruction in the twenty-first century.

Globally, every individual consumes an average of 8 kilos of palm oil in a year[58], with India accounting for 20 per cent of the world's palm oil usage.[59] India is the world's largest importer of palm oil, purchasing over 9 million tonnes every year—approximately two-thirds of our gross edible oil imports.[60]

This means dense, biodiverse rainforests must be felled to meet the demand of human consumption.

What's bad for rainforests is typically bad for humans too.

The quantity of carbon released in the clearing of just one hectare of forest is roughly equivalent to the amount of carbon produced by 530 people flying from Srinagar to Chennai and back in economy class.[61]

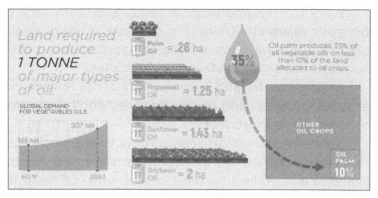

Figure 2-6: Land required to produce palm oil.[62] *Source: International Union for Conservation of Nature*

Sustainable Consumption: For a Sustainable Future

What the palm oil experience teaches us is that just because it's plant-based doesn't necessarily mean it's sustainable.

So how do we go about decoding the complexities of sustainable consumption and incorporate our learnings into a meal plan?

There are two important questions to ask when trying to understand if what is on our plates is good for us and the environment.

The first question is: **how much do you know about what you are consuming?**

Often, as in the case of palm oil, we suffer from what we do not know. Palm oil as a consumable product is not unsustainable, but its production, manufacturing, harvesting and supply mechanisms are.

Want to consume palm oil safely and sustainably? **Do your research.** There is a plethora of companies out there that have committed to sourcing only palm oil produced under **sustainable principles.**

Failing that, try and find suppliers of homemade or artisanal products and ensure that they don't use palm oil to make their products.

However, not all artisanal products are good for the planet, either. Take Instagrammable fad foods like avocado. Riding on the fad-diet wave of the last few years, this fruit is a problem crop that has dire consequences for our planet.

Indian avocado production requires more water than citrus crops that are more suitable to our local climatic conditions. However, the increasing global demand for avocados has displaced citrus crops globally, affecting the livelihood of millions, putting undue strain on our already dwindling supply of fresh water and promoting soil degradation.

This is not to say that avocados themselves are at fault—rather, it is the **carbon footprint** of the fruit that makes it unsustainable for consumption in certain parts of the world. This means that consuming locally grown avocados in Mexico is not unsustainable, but importing Mexican avocados to India is.

Global water shortages in avocado production, among numerous other contributors, mean that excess water must be extracted from groundwater sources and aquifers. Excessive extraction of water from these aquifers has unexpected consequences, such as causing small earthquakes.[63]

Much of the food we eat causes unexpected collateral damage to our environment in much the same way as avocados do. One thing to consider when evaluating your knowledge about food is to find out what the carbon footprint of our meals is—but what does that mean, exactly?

Many of the food items we consume travel over 1,500 miles to reach our plates. This long-distance, large-scale transportation of food accounts for the utilization of large quantities of fossil fuels, which, in turn, lead to rapid GHG emissions. In our current situation, we put almost 10 kcal of fossil fuel-generated energy into every 1 kcal of energy we receive from food.

Calculate your food's carbon footprint[64] here, and use this tool[65] to measure what it would take to offset it.

Crop/animal product	GHG emission (g kg⁻¹)			
	CH_4	N_2O	CO_2	GWP (CO_2 eq.)
Wheat	0.0	0.3	45.0	119.5
Rice	43.0	0.2	75.0	1221.3
Rice, basmati	53.7	0.3	82.5	115.4
Pulse	0.0	0.8	83.3	306.8
Potato	0.0	0.1	10.0	24.9
Cauliflower	0.0	0.1	13.3	28.2
Brinjal	0.0	0.1	12.5	31.1
Oilseed	0.0	1.3	50.0	422.5
Poultry meat	0.0	2.7	50.0	846.5
Mutton [a]	482.5	0.0	0.0	12,062.7
Egg	0.0	2.0	1.0	588.4
Milk [a]	29.2	0.0	0.0	729.2
Banana	0.0	0.2	10.0	71.6
Apple	0.0	1.0	41.7	331.4
Spice	0.0	2.5	100.0	845.0
Fish	25.0	0.3	18.8	718.3

[a] Emission of nitrous oxide and carbon dioxide for milk and mutton production was not considered as buffalo, cattle and goat in India are mostly fed with by-products of crops.

Figure 2-7: Emission of greenhouse gases due to production of various food products from crop and animal.[66] *Source: Research Gate*

So, how do we break the cycle?

This brings us to the next question we must ask ourselves in the quest for sustainable consumption: **how much are we willing to change?**

Changing our consumption habits may seem like a daunting task, but it's important to remember that change is a trickle, not a wave. It occurs in small steps, and every small step counts.

Here are some ways in which we can make a difference that require only minor adjustments to our habits.

Buy Local and Seasonal

One of the most efficient ways to break the vicious cycle of consumption is to **eat and buy local and seasonal food**. This seemingly small step has a ripple effect that can be felt all across the supply chain.

Not only do you decrease your carbon footprint, you also directly benefit small farmers who sell directly to customers. Seasonal foods also fly in the face of unsustainable commercial monocultures, meaning that your food is richer in naturally occurring nutrients and contains none of the harmful chemical fertilizers that one would find in produce that is grown in bulk, all year round.

Urban Farming

If you have switched to working from home over the pandemic, another great hobby to invest in is **urban farming** or the practice of growing your own produce. Garlic, radish, tomatoes, lemons, green chillies and spinach are just a few examples of easy-to-grow greens that do wonders not just for your palate and wallet, but also for the environment.

How, you ask?

Growing our own crops, even if in small quantities in urban homes, has the potential to reduce the energy used to put food on your plate (currently, 25-30 per cent of all greenhouse gases are produced as a result of agricultural production). It also helps keep our waterways healthy and clean by reducing the demand for commercial irrigation and reducing the chemical load entering our waterways from fertilizers and pesticides. Not to mention, urban farming can greatly impact our soil health by minimizing biomass depletion.

Urban farming also directly benefits your health by reducing the chemical quotient of the food you consume, and by creating fresh sources of oxygen via the green plants you will install in your home.

In Ladakh, a majority of people live in homestays, where growing your food is considered the norm. Every house is equipped with a little farm where crops are grown for daily consumption. Very few ingredients are sourced from markets due to the isolated nature of the city, and imports are limited.

Composting

Urban farming also goes hand-in-hand with **composting**, i.e., the sustainable disposal of food waste that can be used to enrich the soil for crops and keep waste out of landfills.

Composting is easier than we imagine. This four-step process can help you get started:

- Collect food scraps and organic waste generated from your kitchen. This can include peels, fruits and veggies, tea bags (minus the metal clip), used coffee grounds and even hair.
- Choose a location for your compost pit. This would ideally be outdoors because, as with most waste, the decomposition process can result in unpleasant smells.
- Colour code for optimal use: Separate your waste into greens, i.e., wet materials like food scraps, and browns, i.e., dry, carbon-rich organic materials such as newspaper, decayed leaves, sawdust etc. Make your compost mix by layering browns and greens alternately while aiming to have a larger ratio of browns to greens. Try and introduce organic catalysers, such as earthworms, which can help break down waste into healthy organic material.
- Wait: Composting takes time, and the time needed can vary depending on your location. A good indicator of healthy compost is the smell—if your compost smells rotten, it's not working. If, on the other hand, it smells earthy and sweet, with an almost pleasant texture, you're on the right track.

Looking to start your own home composting unit? Daily Dump[67] is an organization in Bengaluru that makes its own home composting units that are affordable and extremely easy to use.

Reducing Waste

Reducing household waste, especially when it comes to food, is simpler than we may imagine. It begins in the marketplace—buy loose products in realistic amounts so that there are no extras. Bones produced from meat products and fish, as well as vegetable waste, can be reused to make stock—a healthy and tasty ingredient that goes a long way toward improving the flavour and nutrition profile of your food. What cannot be eaten should be frozen for later use, and food waste can be composted. Besides keeping food waste out of landfills, if the global population managed to stop wasting food altogether, we would completely eradicate 8 per cent of our total emissions.[68]

Packaging

Another invisible source of waste in food is the **packaging**—when shopping, carry a reusable bag, and wherever possible, opt for fruit and vegetables that are packaging-free.

With all of this being said, the best solution to protecting our planet's natural resources while keeping ourselves fed is to change the way we feed ourselves.

So, to go back to the question I had posed at the beginning of this chapter, if we had to draw up a meal plan that protected the health of our planet as well as of all those who inhabit it, what would we eat?

Alternative Meats

While meat consumption is one of the leading causes of climate disruption, there are sustainable alternatives for those with a protein-rich meat diet. Alternative meats and meat products are the perfect solutions for those who want to switch to a more sustainable lifestyle while not compromising on flavour.

Blue Tribe is one such company spearheading the alternative-meat revolution in India. With plant-based alternatives to just about any poultry or cattle product you could think of, Blue Tribe believes in small changes that can have compounding effects over time.

Sohil Wazir, chief commercial officer of Blue Tribe Foods, says:

'Keeping aside the ethical angle of killing another sentient being for a sandwich, the scale at which meat production uses our planet's limited resources is unsustainable. The production of meat alternatives utilizes a fraction of the resources needed for animal meat, and is a necessary step in the eradication of zoonotic diseases like swine flu, avian flu, and more. Technology has now evolved to the level that you can replicate the nutritional qualities of meat, even going a step beyond to address some of the pitfalls of a meat-rich diet, such as high cholesterol. They give you the option to consume what you want to consume, but sustainably, without compromising on taste, texture, or nutrition. Alternative meats don't just save us from our contribution towards rapid climate deterioration, but also our own personal sense of guilt.'

India is home to several other alternative meat start-ups, too, including

1. Good Dot
2. Evo Foods
3. Mister Veg
4. Greenest
5. Vezlay
6. Wakao
7. Imagine Meats

Sustainable Seafood

While overfishing and aquaculture come with their challenges, there is a way to circumvent the ills of the industry while not eschewing seafood entirely. Choose a diverse range of seafood products that are responsibly produced. For example, molluscs like scallops, clams, oysters and mussels feed on microscopic organic matter, including chemical and nutrient-rich agricultural runoff, turning waste into viable carbon stores and a source of food.

Prioritize Plants

With global meat consumption increasing by 58 per cent[69] over the twenty years leading up to 2018, reaching 360 million tonnes, prioritizing vegetables and leafy greens such as kale, spinach and more can make great strides towards an eco-friendly diet. With limited resources needed to produce large quantities of crops, leafy vegetables and green are as good for the environment as they are for us.[70]

Diversify

Did you know that in today's world, 75 per cent of the world's food supply comes from just twelve plants and five animal species?[71] Diversifying our eating habits is not just good for us, but is essential to making our agricultural practices viable in the long term.

Here are some food products that most people ignore that are *actually good* for the environment.

Mushrooms

Mushrooms are fungi, which means that they grow in those odd spaces where most other crops cannot grow. This means that they can also grow in the waste and other by-products that emerge from the manufacturing of other crops. According to this 2017 report by the Mushroom Council, the environmental impact of growing and cultivating mushrooms over two years found that the process utilizes a minuscule amount of water and energy as compared to other crops, with a negligible CO_2 emission rate.[72]

One pound of button mushrooms utilizes only 2 gallons of water to produce, which is a far cry from the average of 50 gallons of water required per pound of other crops.

Mushrooms come in a variety of options, with over 2,000 being edible. They are also great sources of nutrition, containing high quantities of protein and fibre.[73]

Seaweed and Algae

Algae might sound like a strange addition to this list, but it is also responsible for half of all oxygen production on earth, with all aquatic ecosystems being reliant on it for oxygen.[74]

Packed with antioxidants, vitamin C, essential fatty acids and proteins, seaweed and algae can be grown in large quantities underwater without the use of any form of fertilizers or pesticides. The best part? It can be harvested year-round!

Cereals and Grains

Cereals and grains have always been essential to our diets, and are packed with healthy fibres, antioxidants, vitamins and minerals. Low in GHG emissions, cereal grains like wheat require only 138 gallons of water per pound.[75]

The Price of Sustainability: What Are We Missing?

People argue that a completely eco-sensitive plant-based diet is expensive.

While that is true to a certain extent, it is not necessarily accurate.

If you happen to live in India, you might be surprised by how many home-cooked meals are plant-based, with endless potential for sustainability, depending on where and how the individual ingredients are sourced.

Did you know, for example, that all common dals or lentils, including *masoor, arhar, moong* and *urad*, are sustainable, plant-based foods? In fact, all legumes and pulses, like beans, peas, chickpeas, and even *rajma* or kidney beans, are plant-based too.

Legumes and pulses also act as an important part of sustainable cropping systems. They enrich the soil with nitrogen, improving its health and reducing the need for water. They are also resilient crops by nature, with many species even being drought-tolerant, and perfect for cultivation in dry regions.

Sonal Ved, author and digital editor, *Cosmopolitan, Harper Bazaar & Bridal India*, says:

'What you choose to put on your plate can have the most powerful impact on your environmental footprint. At least 3x a day you make choices on how you nourish yourself and these choices can either have a lighter impact on Mother Earth or a heavier one. If healthy, raw or conscious food doesn't appeal to your taste buds, sometimes it is because one has lost touch with their natural food instincts. We are so used to a standard American diet of highly processed, packaged foods. So sometimes, it's about taking a step back and learning what works for our bodies in a way that is not detrimental to the environment.'

Nutritional composition of main millets in comparison to major cereals (@ 12% moisture, per 100 g edible portion).

Nutrients	Finger millet[1;2]	Pearl millet[1;2]	Foxtail millet[1;2]	Proso millet[1;2]	Wheat[1;2]	Rice (white, milled, raw)[1;2]	Rice (brown, medium grain, raw)[3]	Corn grain (white)[3]	Sorghum[3]	Oats[3]	Barley (pearled, raw)[3]
Proximate composition											
Moisture (g)	13.1	12.4	11.2	11.9	12.8	13.7	12.4	10.4	12.4	8.2	10.1
Energy (kcal)	336	361	331	341	346	345	362	365	329	389	352
Protein (g)	7.7	11.6	12.3	12.5	11.8	6.8	7.5	9.4	10.6	16.9	9.9
Fat (g)	1.5	5	4.3	1.1	1.5	0.5	2.7	4.7	3.5	6.9	1.2
Total dietary fiber (g)	11.5	11.3	2.4	-	12.5	4.1	3.4	7.3	6.7	10.6	15.6
Carbohydrate (g)	72.6	67.5	60.9	70.4	71.2	78.2	76.2	74.3	72.1	66.3	77.7
Minerals (g)	2.7	2.3	3.3	1.9	1.5	0.6	-	-	1.6	-	-
Minerals and trace elements											
Calcium (mg)	350	42	31	14	30	10	33	7	13	54	29
Iron (mg)	3.9	8	2.8	0.8	3.5	0.7	1.8	2.7	3.36	4.7	2.5
Magnesium (mg)	137	137	81	153	138	64	143	127	165	177	79
Phosphorus (mg)	283	296	290	206	298	160	264	210	222	523	221
Manganese (mg)	5.94	1.15	0.6	0.6	2.29	0.51	-	-	0.78	-	-
Molybdenum (mg)	0.102	0.069	0.7	-	0.051	0.05	-	-	0.039	-	-
Zinc (mg)	2.3	3.1	2.4	1.4	2.7	1.3	2.02	2.21	1.7	3.97	2.1
Sodium (mg)	11	10.9	4.6	8.2	17.1	-	4	35	2	2	9
Potassium (mg)	408	307	250	113	284	-	268	287	363	429	280
Vitamins											
Thiamine (mg)	0.42	0.33	0.59	0.2	0.45	0.06	0.41	0.39	0.33	0.76	0.19
Riboflavin (mg)	0.19	0.25	0.11	0.18	0.17	0.06	0.04	0.2	0.096	0.14	0.11
Niacin (mg)	1.1	2.3	3.2	2.3	5.5	1.9	4.3	3.6	3.7	0.96	4.6
Total Folic acid (mg)	18.3	45.5	15	-	36.6	8	20	-	20	56	23
Vitamin E (mg)	22	-	-	-	-	-	-	-	0.5	-	0.02

(1) Gopalan et al. (1999).
(2) Gopalan et al. (2004).
(3) USDA National Nutrient Database for Standard Reference, Release 28 (2016).

Figure 2-8: Nutritional composition of main millets in comparison to major cereals. Source: Research Gate

While some naans and other Indian breads use curd and other dairy products, making them unqualified for a plant-based diet, most plain rotis and *parathas* are plant-based too. If you're looking for a sustainable alternative to wheat or *maida,* there are a variety of other flour options that are less harmful to our planet and environment, including amaranth, barley, coconut, chestnut, maize, millet, teff, oat, rye, sorghum, soy and rice flour.[76]

If you're a fan of South Indian food, like I am, you will be pleased to hear that a majority of *vadas*, served with coconut chutney and *sambar*, as well as banana chips, *dosa* and more, are sustainable, plant-based foods as well.

I have travelled across the nation and have never had trouble with my plant-based lifestyle. Here is a visualization of my favourite plant-based foods across India.

Ladakh and Spiti:
Thukpa, vegetarian momos, skyu, thenthuk, timuk

Himachal:
Tudkiya bhath, dhaam, siddu, aktori, madra

Madhya Pradesh:
Poha, kachori, samosa, dal, roti, sabji

Andhra, Tamil Nadu, Kerala, Telangana, Karnataka:
Mysore masala dosa, masala dosa, onion tomato uttapam, idli sambar

Gujarat:
Khaman, Khakhra, thepla, undhiyu, fafda, puran poli

Maharashtra:
Pav bhaji, misal pav, vada pav, sabudana khichdi, pithal bhakri

Rajasthan:
Dal baati, mirchi bada, pyaaz ki kachori, gatte, bajra ki roti with lehsun chutney, aam ki launji, methi bajra puri

The point I am trying to illustrate is that we are asking the wrong questions when it comes to sustainable food consumption. Do not ask why eating sustainably is an expensive undertaking, but rather, ask yourself why dairy products, processed and frozen foods and aerated beverages are so cheap.

What is the *real* cost we are paying?

If a stranger walked up to you and tried to sell you a freshly harvested truffle or a prime cut of Kobe beef for only one rupee, wouldn't you be suspicious?

Now, ask yourself why you do not have the same reaction when it comes to the food you are using for nutrition.

Resources

Here is an inventory of supplementary media material, people who matter and shops in line with the chapter's theme.

Books

1. *Soil Not Oil: Environmental Justice in an Age of Climate Crisis*
    Author: Vandana Shiva
    Published by North Atlantic Books
    https://www.northatlanticbooks.com/shop/soil-not-oil

2. *The Great Climate Robbery*
    Author: Grain Ngo
    Published by New Internationalist
    https://www.amazon.in/Great-Climate-Robbery-System-Drives/dp/9382381686

3. *The Fate of Food: What We'll Eat in a Bigger, Hotter, Smarter World*
    Author: Amanda Little
    Published by Penguin Random House
    https://www.penguinrandomhouse.com/books/536426/the-fate-of-food-by-amanda-little

4. *The Omnivore's Dilemma*
    Author: Michael Pollan
    Published by Bloomsbury Publishing
    https://www.bloomsbury.com/in/omnivores-dilemma-9781408812181

5. *Food and Climate Change Without the Hot Air: Change Your Diet: The Easiest Way to Help Save The Planet*
Author: SL Bridle
Published by UIT Cambridge Ltd.
https://www.uit.co.uk/food-and-climate-change-without-the-hot-air.html

Documentaries

1. *Cowspiracy: The Sustainability Secret*
Directed by Kip Andersen and Keegan Kuhn
Available on Netflix
https://www.cowspiracy.com

2. *Seaspiracy*
Directed by Ali Tabrizi
Available on Netflix
https://www.seaspiracy.org

3. *Kiss the Ground*
Directed by Josh Tickell and Rebecca Harrell Tickell
Available on Amazon Prime
https://kisstheground.com

4. *Earthlings*
Directed by: Shaun Monson
Available on YouTube
https://www.youtube.com/watch?v=8gqwpfEcBjl

5. *What the Health*
    Directed by Kip Andersen and Keegan Kuhn
    Available on Netflix
    https://www.whatthehealthfilm.com

TedTalks

1. How Climate Change Could Make Our Food Less Nutritious
    By Kristie Ebi
    https://www.ted.com/talks/kristie_ebi_how_climate_change_could_make_our_food_less_nutritious

2. Poison on Our Plate
    By Ramanjaneyulu GV
    https://www.youtube.com/watch?v=64RLBgD-Cck

3. Climate Change Is Becoming a Problem You Can Taste
    By Amanda Little
    https://www.ted.com/talks/amanda_little_climate_change_is_becoming_a_problem_you_can_taste

4. Every Argument Against Veganism
    By Ed Winters
    https://www.youtube.com/watch?v=byTxzzztRBU

5. The Fastest Way to Slow Climate Change Now
    By Ilissa Ocko
    https://www.ted.com/talks/ilissa_ocko_the_fastest_way_to_slow_climate_change_now

People

1. Richa Hingle | @veganricha
   Shares plant-based, easy, quick, and nutritious meals for the Indian taste palette
   https://www.instagram.com/veganricha

2. Shweta Dudi | @indvegankitchen
   Plant-Based Cooking
   https://www.instagram.com/indivegankitchen

3. Purvi Shah | @iampurvishah
   Plant-Based Cooking and Food Photography
   https://www.instagram.com/iampurvishah

4. Shivani Singh | @Kahaniwalishivani
   Vegan Recipes and Sustainable Living
   https://www.instagram.com/kahaniwalishivani

5. Vijayalakashmi Vikram | @viji_moo
   Plant-Based Recipes
   https://www.instagram.com/viji_moo

6. Ekta Singh | nurtureyourself_ekta
   Vegan Nutritionist
   https://www.instagram.com/nurtureyourself_ekta

Shops

1. Amala Earth
   Delivers PAN India
   https://amala.earth

2. One Green
   Delivers PAN India
   https://onegreen.in

3. Vegan Dukan
   Delivers PAN India
   https://vegandukan.com/

Organic Grocery

1. Satopradhan Vision
   Delivers PAN India
   https://satopradhan.com/

2. Conscious Food
   Sustainably sourced grains, pulses, lentils, etc.
   Delivers Internationally
   https://www.consciousfood.com

Plant-Based Meat and Alternatives

1. Blue Tribe
   Plant-based Meat
   Delivers PAN India
   https://www.bluetribefoods.com

2. Imagine Meats
   Plant-Based Meat
   Delivers PAN India
   https://www.imaginemeats.com

3. Evo Foods
   Egg Alternative

Delivers PAN India
https://evofoods.in

4. Wakao Foods
    Meat Made from Jackfruit
    Delivers PAN India
    https://www.wakaofoods.com

5. Mister Veg
    Plant-Based Meat, Chicken, and Fish
    Delivers PAN India
    https://www.misterveg.com

Plant-Based Milk and Milk Product Alternatives

1. Alt Foods
    Plant-Based Milk
    Delivers PAN India
    https://www.misterveg.com

2. One Good
    Plant-based Milk, Curd, Ghee, Butter
    Delivers PAN India
    https://onegood.in

3. Soft Spot
    Plant-based Cheese
    Delivers PAN India
    https://softspotfoods.com

4. The Vegan Co
    Specialises in Vegan cheese Dips

Delivers PAN India
https://theveganco.com

5. Bombay Cheese Company
Authentic Vegan Cheese
Delivers PAN India
https://bombaycheesecompany.com

Vegan Cafes

1. Greenr
Gurgaon | Delhi | Mumbai
https://www.instagram.com/begreenr

2. Copper+Cloves Cafe
Indiranagar, Bangalore
https://www.instagram.com/copperandcloves.cafes

3. For Earth's Sake
Galleria Market, Gurgaon
https://forearthssake.co.in

4. GoodDo by GoodDot
Multiple Outlets in Udaipur and Mumbai
https://www.gooddo.co.in/locate-us

5. Just be Resto Cafe
Bangalore
https://www.instagram.com/justbecafe

You can also find a vegan cafe in your city from this link
https://www.veganfirst.com/restaurants

3

Textiles

GHG Emission: 6-8 per cent[1]

This industry impacts the following Sustainable Development Goals (SDGs):
- SDG 1: No Poverty
- SDG 3: Good Health and Well-Being
- SDG 6: Clean Water and Sanitization
- SDG 8: Decent Work and Economic Growth
- SDG 10: Reduced Inequalities
- SDG 12: Responsible Consumption and Production
- SDG 13: Climate Action
- SDG 15: Life Below Water

ACTIVITY

Try and answer the following questions as sincerely as possible:

1. How often do you shop for new clothes?
   - ☐ 1-2 times a year
   - ☐ 3-4 times a year
   - ☐ 5-6 times a year
   - ☐ 7-10 times a year
   - ☐ 10-12 times a year
   - ☐ Others (Please specify) _____

2. How often do you wash your clothes?
   - ☐ At least once a week
   - ☐ More than once a week
   - ☐ Once every few weeks
   - ☐ Others (Please specify)

3. Which of the following fabrics and materials are your garments and clothing made from? Tick all that apply:
   - ☐ Wool
   - ☐ Cotton
   - ☐ Silk
   - ☐ Hemp
   - ☐ Leather
   - ☐ Fur
   - ☐ Khadi
   - ☐ Polyester
   - ☐ Acrylic
   - ☐ Khadi

4. What do you do with your old clothes?
   - ☐ Donate/Hand-me-downs
   - ☐ Upcycling/Downcycling
   - ☐ Throw them away
   - ☐ Don't know what happened to them

5. Where do you shop from?
   - ☐ Online
   - ☐ Retail shops/Fast fashion brands
   - ☐ Local Boutiques/Shops

Once, the bare necessities of roti (food), *kapda* (clothing) and *makaan* (shelter) were considered the essentials for a healthy, happy life.

While a lack of food and shelter still plagues populations worldwide, today, fashion is more accessible than it has ever been before.

With just the click of a few buttons, you can transform everything in your closet, be it a pair of trendy boots, the latest cut of jeans or even a heavily embroidered *lehenga* for your wedding. With prices ranging from Rs 100 to Rs 1,00,000 and above, kapda has become more commodity than necessity for those living in modern metropolises.

Where once the Father of our Nation, Mohandas Karamchand Gandhi, appealed to the Indian public to spin, weave and wear khadi-based clothing as a means to bring about mass employment and eradicate poverty, today, India's relationship with textiles is a far cry from the colonial era.

Today, the textiles and apparel industry in India is the nation's second largest employer, providing direct employment to 45 million people and 100 million people in allied industries. Globally[2], the textile market was valued at USD 993.6 billion in 2021 and is anticipated to grow at a rate of 4.0 per cent annually from 2022 to 2030.[3]

India is home to a massive textile manufacturing hub, with approximately 4.5 crore employed workers, including 35.22 lakh handloom workers across the country.[4] In 2021, Indian textile imports were worth over Rs 11,000 crore out of a total import value of over Rs 29 lakh crore.[5] Surprisingly, this figure represents a decrease in imports from the previous year, which was marked by a global pandemic.

As the adage goes, 'Fashion is a reflection of our times'.

If this is the case, what does the current scope of the fashion industry have to say about the world we live in?

*

If I asked you to guess how many items of clothing I own, what would your answer be? Thirty? Fifty? A hundred?

I have never been much of a fashion enthusiast. As a child, I was limited by my resources. As an adult, I am limited by my interest. To me, the clothes I wear are less a commentary on my personality and more something to keep me warm, cool and protect me from the elements on my adventures in the wild.

Not everyone is like that, and not everyone should be. For some people, the clothes they wear are a part of their job, a means of earning their livelihood and the way they can feed their families. For others, style is a statement, and a way to express their identity. However, there is a darker side to fashion than we may at first realize.

Over my travels, I have had the opportunity to witness first-hand the corruption and evils of the fashion industry, which is killing our natural ecosystems, destroying our climate and harming indigenous communities worldwide.

How, you ask?

Let me tell you the unheard story of the modern fashion industry, and why I, a twenty-six-year-old man who has travelled to the farthest reaches of our country, own only four items of clothing.

Tied up in Knots: The Problem with Textiles

The fourth biggest industry globally, the apparel and textile industry is the second largest source of pollution in the world, beaten only by the oil and gas industry.[6]

Producing 1.2 billion tonnes of CO_2 equivalent emissions (CO_2e) per year, the textile and apparel industry generates more emissions than international flights and maritime shipping.[7]

With new trends hitting the market multiple times every year, and the advent of the age of social media and e-commerce, the demand for fashion has increased exponentially over the last decade.

With increased demand comes the need for increased manufacturing.

As with most other industries, when manufacturing must be upscaled to meet a time-bound demand, corners must be cut and shortcuts must be taken.

Unfortunately, these shortcuts typically mean that someone along the supply chain must lose so the industry can win.

In the case of the fashion industry, that someone is all of us—we just don't know it yet.

Not only does this accelerated cycle of demand and supply put monumental stress on our natural resources, but it also adds to energy costs and results in harmful chemicals being introduced into our air,

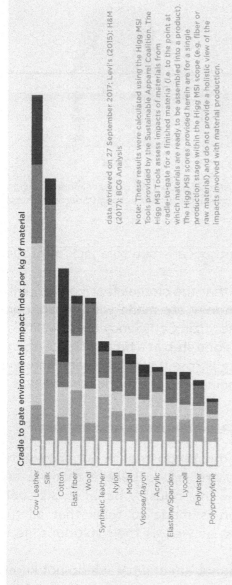

Figure 3-1: Cradle to gate Environmental impact index per kg of material.[8] Source: International Labour Organization

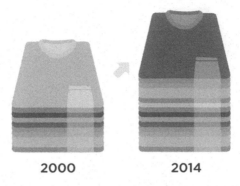

2000   **2014**

Figure 3-2: An average customer bought 60 per cent more clothing in 2014 than in 2000. Each garment was half as long.[9] Source: McKinsey and Company

land, lakes, rivers and oceans due to a lack of proper waste management.

This also means that the clothes that you buy with your hard-earned money are made with cheap, non-durable material, prioritizing efficiency over quality.

But let's take this one step at a time and analyse the gaps in sustainability across the textile manufacturing supply chain—beginning with the fabrics themselves.

Wool Over Our Eyes: Manufacturing Fabrics

As is the case with most industries in our capitalistic society, consumers are led to believe that we are responsible for how the industry behaves. As such, the average person is made to think that all the industry's problems, as well as their solutions, lie in their hands.

However, we cannot affect what we do not know. While our consumption patterns do indeed affect the industry, there are many hidden factors along the supply chain that we, as consumers, are not aware of.

While some may willingly turn a blind eye, the issue comes from an industry-wide practice of gatekeeping information, so that what benefits the industry can be kept alive, even if it might mean harming someone else.

Let's examine some of these issues from the lens of the materials and fabrics themselves and explore the many ways in which the demand for them fuels hidden unsustainable and exploitative practices, which

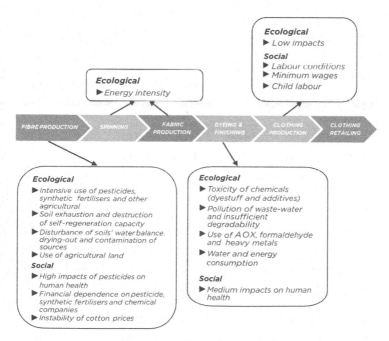

Figure 3-3 Environmental and social impact of manufacturing [10] *Source: Environmental Analysis of Textile Value Chain: An Overview*

S.No.	Commodity	Apr-Mar 2019	Apr-Mar 2020	%Growth	%Share
1	MANMADE YARN, FABRICS, MADEUPS	2,202.13	2,190.72	-0.52	0.46
2	COTTON, RAW INCLD. WASTE	633.05	1,328.41	109.84	0.2
3	OTH TXTL YRN, FBRIC MDIP ARTCL	913.62	860.58	-5.8	0.18
4	COTTON FABRICS, MADEUPS ETC.	497.94	551.81	10.82	0.12
5	RMG COTTON INCL ACCESSORIES	547.74	534.05	-2.5	0.11
6	MANMADE STAPLE FIBRE	467.38	491.46	5.15	0.1
7	RMG MANMADE FIBRES	323.98	356.73	-10.11	0.08
8	RMG OF OTHR TEXTILE MATRL	208.63	231.53	10.98	0.05
9	WOOL, RAW	310.3	225.31	-27.39	0.05
10	SILK, RAW	148.38	162.38	9.44	0.03

Top 10 Export Commodities

Values in Million USD

S.No.	Commodity	Apr-Mar 2019	Apr-Mar 2020	%Growth	%Share
1	RMG COTTON INCL ACCESSORIES	8,694.73	8,642.88	-0.6	2.76
2	COTTON, FABRICS, MADEUPS ETC.	5,947.20	5,967.02	0.33	1.91
3	MANMADE YARN FABRICS, MADEUPS	4,980.51	4,820.97	-3.2	1.54
4	RMG MANMADE FIBRES	3,852.96	3,506.00	-9.01	1.12
5	RMG OF OTHR TEXTILE MATRL	3,222.69	3,064.92	-4.9	0.98
6	COTTON YARN	3,895.52	2,760.51	-29.14	0.88
7	CARPET (EXCL. SILK) HANDMADE	1,465.74	1,353.00	-7.69	0.43
8	COTTON RAW INCLD. WASTE	2,104.41	1,057.82	-49.73	0.34
9	MANMADE STAPLE FIBRE	570.81	503.04	-11.87	0.16
10	OYH TXTL YRN, FBRIC MDUP ARTCL	457.93	477.28	4.23	0.15

Table 3-1: Top 10 export commodities.[11] Source: Government of India

Processing subcategory	Water consumption (m^3/ton fibre material)	
	Minimum	Maximum
Wool	111	285
Woven	5	114
Knit	20	84
Carpet	8.3	47
Stock/yarn	3.3	100
Nonwoven	2.5	40
Felted fabric finishing	33	213

Table 3-2: Average water consumption for various types of fabrics.[12]
Source: Environmental Analysis of Textile Value Chain: An Overview

are key to the current state of fashion, beginning with the most popular fabric of all—cotton.

Cotton has been the fibre of choice to clothe humanity for era upon era. Today, it is notably the most common material used in garment production.[13]

Conventional cotton farming begins with genetically modified seeds, which require various resources, including water, energy, vast expanses of land, pesticides and fertilizers to bring to harvest. These water and fertilizer-intensive crops take a toll not only on the limited natural resources we have but also on the fragile integrity of the soil upon which they are grown. According to estimates, a kilogram of textiles utilizes approximately 50–100 litres of water.

If breezy cotton isn't your style, perhaps you prefer fabrics sourced from animals, such as wool, cashmere, leather, fur or silk.

If you, like me, grew up imagining wool production to be just like it is in those old-time cartoons—a fluffy sheep being sheared with a delicate hand by an endearing farmer humming the words to 'Old McDonald'—I am sorry to say that the world has moved far, far away from that humble picture.

A majority of fabrics sourced from animals hide a dark history of abuse, mistreatment and gratuitous culling behind their exotic, expensive exterior. Every year, billions of animals must endure the horrors of life on a factory farm, where, after years of torture and torment, they are killed for their skin, fur or feathers.

Today, animals such as sheep, cows and goats are housed in unsanitary, unethical conditions, tortured, fed growth hormones and even killed for their skin and fur.[14]

In the natural world, sheep produce just about enough wool to protect themselves from the elements. However, human beings selectively breed them to produce wool in commercial quantities.[15]

At factory farms, speed and efficiency are prioritized above all else. Workers are typically paid by volume rather than by the hour, leading to them treating animals in rash, unsympathetic ways to expedite the process. This means that shearers are encouraged to work fast, violently shearing wool and leaving the frightened sheep with deep cuts and gouges that are hastily sewn together to get to the next one in line. As PETA revealed, workers were also caught giving the animals painkillers before hastily shearing them, so that they would not fight back.[16]

After a lifetime of stolen autonomy and daily torment, these animals are no longer of use to the industry and are disposed of in the most cost-efficient way possible. This often entails violent acts of slaughter, including bludgeoning, anally-induced electrocution and lethal gassing. In some cases, animals may be skinned alive or dismembered so that their parts may be used to the most optimal and profitable degree.

While the treatment of sheep in the wool industry may be a relatively well-hidden secret, the same simply cannot be said for leather. By definition, leather is dead, treated skin, and so the act of buying leather is in direct support of animal cruelty and slaughter.

In India, it is punishable by law to kill healthy, young cattle. To circumvent this law and turn a profit in times of low milk production, farmers will sometimes deliberately maim cattle, breaking their limbs or feeding them poisons so that they may be declared fit for the kill.

A common source of leather for the so-called *'ahinsak'* footwear, which is marketed to be cruelty-free and respectful of animals, is the dairy industry. Male calves, old cattle, and more who serve no use to dairy farmers are sold for slaughter or are starved to death or maimed so that their skin may be converted into ahinsak footwear.[17]

According to Taran Chhabra, co-founder of Neeman's Shoes:

'To me, sustainable fashion is that which is made consciously, calculatedly, and lasts long. For that to happen, not only is there a need for responsible consumption, but also purposeful and ethical production. More than 21 billion pairs of footwear are manufactured annually. Footwear industries produce the most significant quantity of leather waste, from leather trimmings to shavings and leather dust. The majority of the waste, however, is generated from post-consumer footwear waste,

> i.e. end of life of footwear which mainly goes into landfills. Chemical adhesives and tanning chemicals are used to preserve the materials, such as leather, in shoes. These chemicals, when introduced into the environment through discharge from factories, harm the wildlife and the people living nearby.'

In the case of exotic furs, trapped animals suffer greatly from loss of blood, dehydration and gangrene from infected wounds, and can even become paralysed by shock. They remain in this state for days on end, and are often attacked by predators before trappers return to kill them.[18]

If the cruelty meted out to these living creatures isn't enough, the commercial production of these fabrics has a host of other detrimental effects on the environment.

Soil bears the brunt of the adverse effects, undergoing compaction due to the hooves of sheep and goats. Deforestation for feed and land clearing for grazing, as well as overgrazing, can result in poor soil health and a loss of biodiversity. The extensive use of chemicals to ward off parasites can also lead to soil contamination, robbing it of its essential nutrients.

Moreover, the ruminant animal husbandry industry results in the generation of harmful quantities of methane, which has a CO_2 equivalent of 84x.

Even silk, which is sourced from commercial silkworm production, involves the extensive use of chemical fertilizers and pesticides for the cultivation of mulberry trees. Mulberry leaves are the only leaves that silkworms consume, and the leaves are required

Category	GWP (kg CO_{eq}/kg)	Renewable CED (MJ/kg)	Nonrenewable CED (MJ/kg)	Eco-toxicity (CTU/kg)	ALO (m²a/kg)	ULO (m²a/kg)	BWF (m³/kg)	FE (gP_{eq}/kg)
Farm practices	80.9	1613.6	244.4	1043.1	35.6	1.37	54.0	3.0
Recommended practices	52.5	1350.6	116.7	522.9	19.8	1.13	26.7	4.8
Cotton	3.4	19.7	0.1	71.2	7.8	0.02	7.0	0.8
Nylon 66	8.0	1.3	7×10^{-4}	6×10^{-4}	2×10^{-4}	4×10^{-4}	0.2	0.3
Wool	18.5	81.7	0.1	3.4	53.5	0.36	0.2	0.5

ALO, agricultural land organisation; BWF, blue water footprint; CED, cumulative energy demand; FE, freshwater euphoric action; GWP, global warming potential; ULO, urban land occupation

Table 3-3: Environmental impact of silk compared to other fabrics.[19] Source: The Textile Institute of Book Series

to be fresh as silkworms source all their moisture from the leaves.

Silk production is also a water-intensive process, with 66.66 litres of water needed for just one piece of handwoven silk, not to mention the 120 litres of water needed in the degumming process to remove silk glue.[20]

We all know that most commercial products that involve animals typically also involve some form of cruelty towards them, but what about acrylics and synthetic clothing? Surely they can't be as bad, right?

Synthetic fabrics, such as acrylics and polyester, account for more than half the world's garments. However, in terms of their environmental impact, acrylics, polyester and other synthetic materials are probably the most harmful of them all.

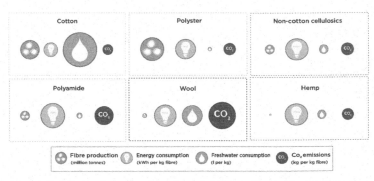

Figure 3-4: Polyesters are more harmful than other materials.[21] Source: Nature Reviews Earth and Environment Volume 1

Made using non-renewable resources, such as fossil fuels, these fabrics are non-biodegradable, emit toxic chemicals and GHGs during production, and pose several health issues for humans and animals alike.

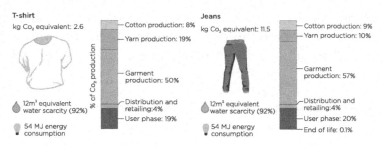

Figure 3-5: Energy taken by various fibre to process.[22] Source: Nature Reviews Earth and Environment

The production of 1 kg of acrylic fibre requires 157 megajoules of energy and 210 litres of water and leads to 5 kg of carbon dioxide emissions. It also leads to the production of microplastics. Plus, they aren't recyclable.[23]

Dressed to Kill: Processing and Finishing

Now that we have discussed the lack of sustainable manufacturing of fabrics and materials, let's skip to the next stage: processing.

The processing of fibres to produce fabric is taxing to our ecological health.

Individual fibres must be treated using harmful chemicals mixed with litres upon litres of water. The machinery utilized in the process also makes use of harmful chemicals like lubricants, plastics and chemicals for treatment. These lead to the generation of particulate matter, water-based effluents, volatile organic compounds (VOCs) and more, which negatively impact our health in the long term.

The processing stage produces effluents in momentous quantities. The handling of these materials is an equally challenging undertaking, which makes use of non-biodegradable chemicals, hydrocarbon-based conditioners, synthetic compounds and several colour-fixing agents.

Once the fibres have been processed into fabrics, they must be finished, i.e., made ready to sell.

This process typically involves bleaching, dyeing, printing and more, each of which utilizes harmful, toxic chemicals and heavy metals that then bleed into water. Textile effluents are also marked by a high pH level, with high quantities of dissolved solids like sulphur and salt. They are also of higher temperatures than the water bodies they are dumped into, leading to lower levels of available oxygen for flora and fauna, and overall disastrous consequences for surrounding areas and communities.

The waste generated from this process in the form of emissions is harmful to air and water quality, with a large amount of chemically polluted solid waste being generated too.

Hanging by a Thread

Unsustainable Practices on a Micro Scale

Microfibres are the silent killers of our times and have only recently been taken seriously by the world population.

We have already spoken about the unsustainability of synthetic clothing, but their impact can mean more than meets the eye. Marketed as being both durable and affordable, and having exponentially increased in popularity since the 1960s, synthetic garments are

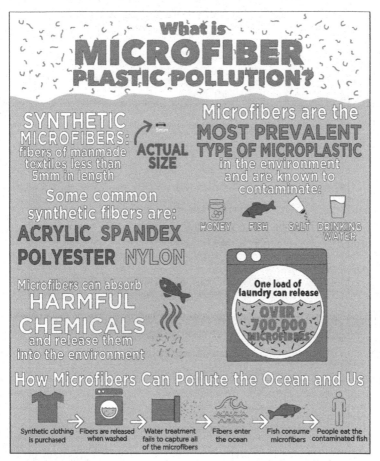

Figure 3-6: Understanding micro-fibre pollution.[24] *Source: Save Our Shores*

major contributors to the production of microfibre, the most prevalent type of microplastics infiltrating our natural world today.

Microplastics are classified as plastic pieces less than 5 mm in diameter. Much like oxygen, plastic microfibres are all around us, even if we cannot see them.

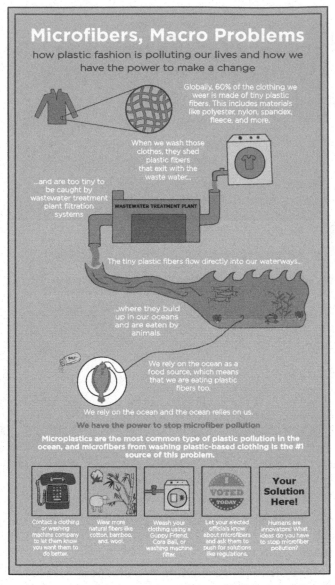

Figure 3-7: How plastic fashion is damaging our life.[25] Source: Microfibres, Macro Problems

Every time a synthetic garment is washed, about 1,900 individual microfibers are released into the water, making their way into our oceans, from there into the fish and other organisms that reside there, and eventually into our food chain.[26]

Being non-biodegradable by nature and minuscule in size, these microfibres are easily washed into neighbouring water bodies, and can also be found in freshwater systems worldwide.

Quantification for the Indian Ocean revealed that there are approximately 4 billion microfibres per square kilometre. In fact, as much as 20 per cent to 35 per cent of all primary source microplastics in the marine environment are from synthetic clothing, according to academic estimates.[27]

A Shade Darker:
Exploitation and Effluents in Coloured Fabrics

If there is one thing we have learned from years of debates surrounding global warming and climate change, it is that developing and underdeveloped nations often bear the brunt of the actions of the developed world.

This is also true in the case of fashion and apparel. The fashion industry is notorious for human exploitation, particularly in the case of poor countries.

Due to the availability of cheap labour, countries such as India, China, Bangladesh and Indonesia, among others, are manufacturing hubs for textile companies that serve the demands of the developed nations. These countries are highly dependent on coal and

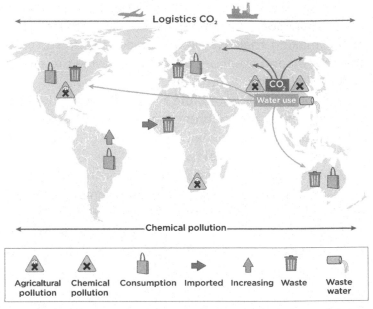

Figure 3-8: Because of cheap labour, developing nations serve as manufacturing hubs for fabrics.[28] Source: Nature Review Earth and Environment

other non-renewable resources and lack proper waste management capabilities.

In our own country, major rivers like the Kasadi in Maharashtra and the Yamuna have undergone rapid acidification due to wastewater dumping and the influx of toxic chemicals into the water systems.

Bangladesh's capital city of Dhaka is home to 719 washing, dyeing and finishing factories, which produce a whopping 200 metric tonnes of wastewater per tonne of fabric produced. This wastewater is unloaded into four major local rivers, which supply water to 18 million residents. Despite the existence of the four rivers, the area's water supply is currently under serious threat due to extreme levels of pollution.[29]

China, the world's largest exporter of clothes, has suffered greatly due to the development of its textile sector. With rapid growth witnessed across China's industrial sector, the textiles market rapidly evolved to keep up with the increasing global demand, resulting in massive environmental costs. As compared to the rest of the world, China's per capita water availability is only a quarter of the world's average. The Chinese textile industry and its associated water pollution only add to local water shortage issues, exacerbating public health issues and catalysing the destruction of its already fragile ecosystem.

The Pearl river, China's third longest river, was reported to have become so polluted that a majority of its tributaries were classified as falling below the lowest national surface water quality standard, and unfit as a source of drinking water.[30]

Despite the passing of several laws and regulations, and strict warnings delivered to textile manufacturers, the dumping of untreated waste did not stop. Some state that one can get a sneak peek of the colour of the season simply by observing the colour of the rivers in China.

According to research conducted by the World Bank, approximately 17-20 per cent of global water contamination figures are the exclusive result of colouring and finishing processes in the textile industry. Additionally, 72 distinctive poisonous synthetics found in water are the direct result of colouring and dyeing.[31]

The colours used in the dyeing process are, by nature of their function, vibrant, deep and easily noticed by the human eye. When effluents from dyeing and manufacturing are released into surrounding water bodies, the visibility levels of the water itself is reduced. This low visibility spells doom for the water bodies' flora and fauna.

Due to the varying degrees of refraction associated with different coloured dyes, sunlight cannot penetrate the coloured water, greatly affecting natural processes such as photosynthesis and algae production, that sustain both plants and creatures native to the water systems.

One of the worst known cases of textile-related pollution is Indonesia's Citarum river. With over 200 textile factories situated along the riverbank, the water body has been officially labelled the most polluted river in the world and is completely uninhabitable for fish and other animals.

Over 60 per cent of the clothing manufactured in Indonesia is exported for sale to other countries, and the locals are left to deal with the consequences. The mercury levels in the Citarum are reported to be more than 100 times the accepted standard. The salts and other dissolved solids in the water make them highly unstable and unsuitable for both domestic use and irrigation.

However, residents living along the riverbank are entirely dependent on it for daily consumption and agrarian practices. According to surveys, 9 million people are currently living in direct or close contact with the river. This is especially troubling, given that research has shown that the levels of faecal coliform bacteria in the water are more than 5,000 times the mandatory limit. With no viable alternatives, the locals are forced to use the water in its most polluted state.[32]

Recent studies have detected pollutants released by the industry along every stage of crop production, and even in the bodies of the residents. This is because effluents from textile factories also contaminate surrounding groundwater reserves. When treated or untreated wastewater is used for irrigation, effluents pollute the soil and diminish its integrity greatly, making it less permeable and more resistant to agrarian activities like ploughing and germination. Heavy metals found in run-off from textile factories also bind with soil particles, making their way into crops and the food chain.

These pollutants have also been linked to health problems, including irritation of the skin and lungs, as well as high rates of cancer, chronic diseases such as tuberculosis and congenital disabilities, detected in communities living near highly polluted rivers and textile factory water outlets. The contaminants being pumped into the waterways are mingling and mixing, creating complex mixtures that grow even more complex every day, and pose an unknown and unquantifiable risk to the environment.

Besides the detrimental effects on the local population, the textile industry's worst victims are, more often than not, those who should benefit from it most directly—the labourers.

In China, it was famously discovered that the persecuted Uyghurs and ethnic Kazakhs who were imprisoned within the internment camp system were being used as forced labour in the manufacturing of textiles for fast fashion brands.[33]

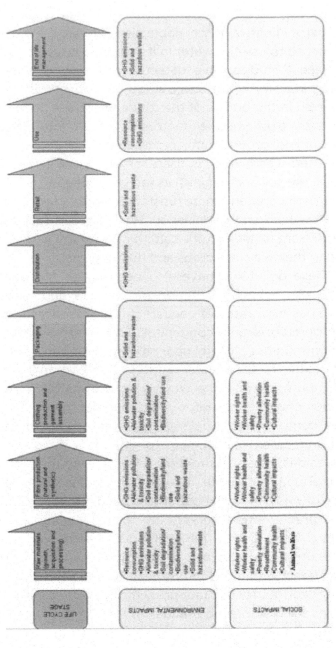

Figure 3-9: Sustainable clothing action plan.[34] Sources: Government of United Kingdom

Another shocking find from the International Labour Organisation declared that 'Approximately 170 million are engaged in child labour, with many making textiles and garments around the world'.[35]

While in developing countries, the clothing and textile industry is seen as a stimulant for economic growth, it comes with more troubles than the rosy picture this notion implies.

While the industry does employ people at mass scales, it hinders their social mobility. The skill development of workers is siloed and of lower value to the overall supply chain, meaning that workers are trapped in the same positions for years on end with no way out. They also have limited access to the market, with controlled information channels and a lack of clarity on trade terms.

Labour in these developing countries typically suffers from poor working conditions, operating in sweatshops with low wages, long hours with limited breaks, and a lack of workers' rights and sufficient equipment to protect themselves from health and safety risks associated with chemical exposure and the operation of heavy machinery.

So, next time you're doing some light shopping for your next night out, ask yourself—is that new pair of jeans worth the human rights crisis you are unwittingly propagating?

Fast Fashion, Slow Decay

The emphasis on profits and efficiency in the digital age of fashion and apparel led to the emergence of a new era for textiles—one defined by what is known as 'fast fashion'.

Fast fashion is a phenomenon born from consumption patterns, both genuine and manufactured. Fashion is defined by the season, and with every season comes a new style—or rather, multiple styles.

TRADITIONAL: 2 CYCLES PER YEAR

TYPICAL FAST FASHION: 50 CYCLES PER YEAR

Figure 3-10: Traditional vs fast fashion.[36] *Source: World Resource Institute*

As the technological revolution democratized our access to the global market, the demand to keep up with international trends grew as well. To meet this growing demand, large-scale retailers optimized their manufacturing processes to produce low-priced, style-appropriate garments designed to move from design to retail within the least possible time.

Over the past three years, the movement of styles from fashion shows to consumers has increased by 21 per cent.[37]

This trend of keeping up with the times also means that the clothes that were considered fashionable a week ago will be obsolete by next week. As a consumer, this means there is always more to be bought, more to be worn and more to be discarded—which is exactly what the industry wants.

This use-and-throw mentality is one of the main contributors to the pollution and exploitation associated with the textile industry. In 2021, the apparent consumption of apparel worldwide was approximately 168.4 billion pieces.[38] According to estimates, this value is expected to increase in the coming years to 197.3 billion pieces in 2026. Over the last fifteen years, worldwide clothing production (and consumption) has doubled.[39]

Social media has accelerated this process, functioning as a digital runway for future styles and seasonal trends. Instead of new stocks every season, retailers are now running new stocks every two weeks.

It has been estimated that there are twenty new garments manufactured per person each year. In a world of over 7 billion people, this number can spell doom for our planet.[40]

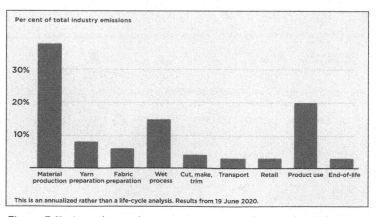

Figure 3-11: Annual greenhouse gas emission of apparel and footwear brand.[41] Source: McKinsey and Co

Disposal: Out of Sight, out of Mind

Have you ever thought to ask yourself, what happens when you throw out old clothes?

Well, if you live in a developed nation, the answer is—they become someone else's problem.

The global south acts as a dumping ground for the north, not only in terms of waste products generated during the manufacturing of the fabrics but also solid waste generated from discarded garments.

In Africa, textile waste imported in the form of second-hand clothing from developed nations has its own term in the local tongue. The Kiswahili word '*mitumba*', i.e., bale or bundle, refers to the practice of selling to local retailers by the bundle. Unfortunately, the sheer quantities being imported into the area far

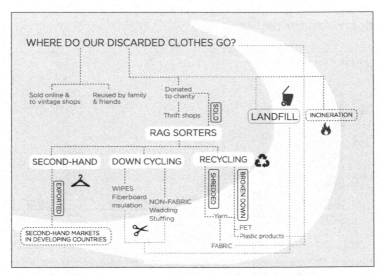

Figure 3-12: The route taken by discarded clothes.[42] Source: Cooper Hewitt infographic

exceed their value in the marketplace, resulting in a massive portion of the fabrics being dumped into landfills and natural landscapes.[43]

However, there is another way in which the textile industry is coming back to haunt us—one we cannot simply ship away to another country's landfills.

Window Dressing: What Are We Missing?

If all of this information is starting to make you feel guilty for the pain and suffering the textile industry causes to millions upon millions of people and living things year upon year, it might be a good time to pause and take a breath.

I can say with confidence that as consumers, and as citizens of the world, we do not mean the harm we inadvertently cause. However, as the saying goes, the road to hell is paved with good intentions.

But Aakash, I hear you ask, *how are we to know the ripple effect a seemingly innocent purchase can make? How are we at fault?*

The answer is that *we are not.* Capitalistic societal structures are built to make a profit and will do just that, no matter the cost to others.

One of the most pertinent examples of this in action is 'greenwashing'—a marketing tactic used by big industry players with the express intention of duping consumers into believing their choices are ethical and sustainable, without having to take accountability for the problems along their supply chain and manufacturing process.

In the not-so-distant past, a well-known Swedish multinational clothing company, a cultural symbol

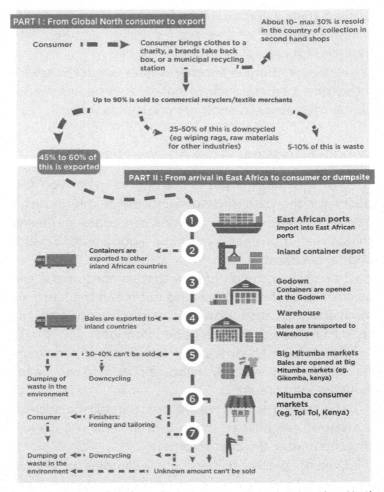

Figure 3-13: Flow of clothes and clothes waste from global north to North Africa.[44] Source: GreenPeace

and one of the most beloved clothing brands for teenagers and young adults, was widely applauded for its recycling program.

The program allowed customers to bring in their old clothes to be recycled into brand-new products.

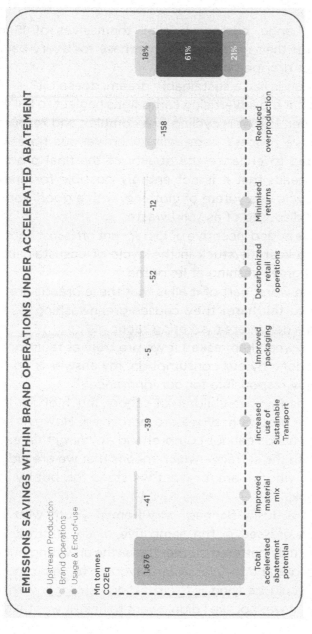

Figure 3-14: Emissions savings with brand operations under accelerate abatement.[45] Source: Global Fashion Agenda

In exchange, customers avail themselves of 15 per cent off their next in-store purchase for every bag of textiles dropped off.

Sounds like a sustainable dream, doesn't it?

What the advertising campaigns neglect to tell you, however, is that recycling is a complex and resource-intensive process, especially when various fibres are blended to enhance the quality of the final product. This means that it is not entirely possible for brands to recycle every item of clothing, with a good portion being disposed of as solid waste.

The added incentive of 15 per cent off is just another tool to keep us stuck in the cycle of unsustainability and is greenwashing at its prime.

The worst part of it all is that these brands are not blind to the harm they cause—greenwashing is very much a deliberate marketing tactic.

So, when I am asked if we are truly at fault for the harm done by our consumption, my answer is no. We are only responsible for our ignorance.

When we are children at school, much of our lives revolve around how we are perceived. However, our young minds cannot comprehend anything that exists beneath the surface—which means that we are judged not by who we are, or what we stand for, but by how we look.

As a dusky-skinned, wiry-framed youth who was always dressed in the same five or six ill-fitted, off-brand, ugly pattered clothes, I was the unwitting butt of every joke and the classroom favourite to be picked on.

I would be lying if I said I didn't struggle with my image. It was not like I didn't want to fit in. It was that, for reasons beyond my control, there was little I could do to change other people's perception of me. I vividly recall standing in front of the mirror, trying on combination after combination, negotiating with myself in a hopeless bid to look like one of the so-called cool kids.

But why am I telling you this?

As an adult, I understand the value of utility over style—but the core driver of profit in the world of fashion is this exploitation of perception. The industry exploits our self-image to make a profit, and to keep us on a constant search for a style that fits our personality. However, the kicker is that our personalities, much like every other part of us, are ever-changing. What truly matters is how we perceive ourselves—and this is not a matter of what we wear, but who we are.

With the advent of the age of social media, and with e-commerce apparel brands promising same-day deliveries on the latest trends, we must understand that there is a billion-dollar industry at work in the shadows, tailoring their every move to get you to click that button and swipe that card.

Solutions: A Stitch in Time

Owning four pairs of clothes is not a style statement, but it is a statement in and of itself.

To illustrate, let's do some quick maths.

If you have five shirts and four pairs of pants, how many different outfits can you put together?

The least number of combinations one can come up with is twenty. If I pair a shirt and a T-shirt together, that number rises to about 100. If I go shirtless while at home, that number rises exponentially.

GARMENT LIFE STAGES

POTENTIAL LEVERS

Garment Life Stages		
Upstream value chain	**Raw material production**	(1) Decarbonized material production Improvements across the production & cultivation of key materials e.g. Cotton, Polyester & Viscose	(2) Improved material mix Decarbonization through improved mix of alternatives for existing materials and introduction of new materials
	Material preparation and processing	(3) Decarbonized material processing Improvements in energy mix and efficiency in processing	(4) Minimized production wastage Reducing waste generated in the processing stages
	Product manufacturing	(5) Decarbonized garment manufacturing Improvements in energy mix and efficiency across manufacturing countries	(6) Minimized manufacturing wastage Reducing waste generated in the manufacturing stages
Transport and distribution	**Sustainable transport**	(7) Increased use of sustainable transport Improvements in fuel mix, energy efficiency of fleets and operational improvements	(8) Improved packaging (manufacturing through retail) Decarbonization through carbon friendly material mix and reduction of packaging usage
Retail		(9) Decarbonized retail operations Improvements in energy mix and efficiency across retail	(11) Reduced overproduction Reducing waste generated due to unsold stock in retail
	Operations	(10) Minimised returns Decarbonization by limiting retail returns (retailer and consumer)	
	New business models	(12) Increased use of rentals models Promotion of subscriptions or one-time rental offerings	(14) Introduction of refurbished/upcycled products Promotion of re-furbished/upcycled product offerings
		(13) Increased use of re-commerce models Promotion of 2nd-hand sales (direct or through platforms)	(15) Introduction of product repair services Promotion of repair services to extend product life
Product use	**Washing and drying**	(16) Reduced washing and drying Reduced washes and improved care (e.g. temperature selection)	
End-of-use	**Recycling and collection**	(17) Increased recycling & collections Increased recycling and collections to minimize incineration without recovery and landfill	

This is all while ignoring shoes, socks, underwear and more.

If I have a hundred outfits for 365 days in a year, I am likely to repeat my outfits one or two times. That doesn't seem so bad to me.

However, you are not like me, and you shouldn't have to be. There are numerous other ways to clothe yourself that do not involve a highly minimalistic way of living, and that, in the long run, might just be the answer to our global climate crisis.[46]

Let's look at just some of these in further detail.

Sustainable Fashion: Conscious Clothing

Sustainable fashion refers to apparel manufacturers that advocate for sustainability not just in their materials, but along every stage of the manufacturing process. These clothes are non-resource-intensive, higher-quality, skin-friendly, and just as stylish as anything you would buy at fast-fashion retailers.

Sustainable fashion takes root in sustainable cultivation, which stems from organic farming practices that eschew typically preferred pesticides, fertilizers and harmful chemicals through the manufacturing process.

For example, hemp, a natural plant fibre, is more sustainable than water-intensive cotton and can go a long way toward correcting some of the unsustainable practices along the supply chain of textiles.

Hemp produces 250 per cent more fibre than cotton and 600 per cent more fibre than flax on the same land and has the highest yield per acre of any natural fibre. Adding to this, hemp is a resilient crop, able to grow rapidly and in a variety of climatic conditions. In fact, hemp plants grow so quickly that they have been known to overgrow other weeds, and require no pesticides, fertilizers or genetically modified seeds to grow to maturity.[47]

According to Yash Kotak, co-founder of Bombay Hemp and Co:

> 'Hemp is a powerhouse of sustainability through its entire lifecycle. Absorbing almost 15,000 tonnes of CO_2 per hectare, hemp enriches the soil, making it more fertile. It also retains colour dyes better than cotton, increasing its usability. Interestingly, hemp is also resistant to UV rays—it blocks more than 50% of UV rays compared to conventional cotton garments— the perfect ingredients for slow fashion. Recent years have seen such alternative fabrics become less expensive, and more accessible to people from all walks of life, due to an increase in government and industrial support for manufacturers, which, to me, is a step in the right direction!'

Adding to its resilient nature, hemp plants require little water, eliminating the need for resource-intensive irrigation, and have been proven to benefit the soil they are cultivated in by replenishing vital natural nutrients, extracting pollutants such as zinc and mercury and minimizing soil erosion via their root system.

Hemp cultivation also prevents the harm done by monocultures, being a favourite rotation crop between harvest cycles to replenish the soil's nutritional content.

Bamboo is another sustainable alternative to fabrics. Being known as a 'miracle product' and an 'eco-crop', bamboo is a fast-growing grass with over 1,000 varieties all across the globe. A single bamboo plant can grow up to four whole feet in a single day and releases 35 per cent more oxygen into our atmosphere than other similar crops. A fast-growing and resilient crop, bamboo plants mature within seven years and rarely need to be replanted due to their thriving root network, which sprouts new shoots on an almost continuous basis.[48]

While the process of converting the harvested crop to fibres has stirred up controversy among sustainability advocates, its light, flowy texture has made it a runway favourite for sustainable fashion showcases.

Another homegrown example of sustainably produced conscious clothing is one that we Indians have been aware of since the freedom struggle: khadi.

The traditional production of khadi makes use of only a spinning wheel and no machines or external energy sources. This means that a khadi garment has a lower carbon footprint even when compared to most other sustainably produced clothing. Khadi is also a water-efficient fabric, with one metre utilizing about 3 litres of water, as compared to the 55 litres of water needed to produce one metre of mill-produced fabric.[49]

Due to its human-centric process, khadi clusters also provide employment and a source of income to indigenous and impoverished communities in rural areas of our nation, as well as in other nations like Mexico.

This shift in consumer mindset towards consciously produced apparel is already picking up steam. In fact, the value of the ethical clothing market increased by 19.9 per cent in 2018, according to Ethical Consumer magazine.[50]

Luxury fashion designer Rahul Mishra says:

'No resource or raw material is harmful to the planet if it's used within limits. For instance, 'plastic' being used as a core material for a luxury jacket, only one copy of which is to be made, with the intention of remaining in someone's wardrobe for decades, would be far better than making thousands of disposable t-shirts out of organic cotton. It is not only important for the consumer to invest in environmentally conscious clothing, but also to purchase fashion that holds longevity and relevance in their wardrobe for a longer time. What's harmful to the planet is the exploitation of any resource and the lack of value that a piece of clothing may hold.'

One great example of a sustainable clothing brand is AirInk. With a sustainable alternative to dyes, composed of condensed carbon-based gaseous effluents generated by air pollution due to the incomplete combustion of fossil fuels, AirInk is reversing the impact the textile industry has on our planet while still providing us with a product that is durable, reliable and stylish.

Another example is Graviky Labs, an MIT spinoff working on developing cleantech solutions. Within a year, Graviky labs, led by Anirudh Sharma, had captured

1.6 billion micrograms of particulate matter, which equates to cleaning 1.6 trillion litres of breathable air.[51]

There are a host of other conscious fabric options for one to explore. These include Econyl, Lyocell, Modal, EcoVero, Piñatex, Woocoa, Cupro, Qmilk and more. Nowadays, it is also possible to buy clothes with a Sustainable Clothing Certification to guarantee you are making the most conscious decision for your wardrobe. Look for garments with chemical content certification labels such as OEKO-TEX®, GOTS, or BLUESIGN®.

Perhaps the most important factor to discuss while talking about conscious clothing is that any discussion about sustainable fabrics promotes its need to the consumer as well as to the manufacturer. As consumer preferences slowly shift towards more ethically produced fabrics, the industry will follow suit, setting off a chain that affects not just the individual, but the entire collective.

Founder of No Nasties Apurva Kothari says:

'You have to see the true cost of what you are buying in fast fashion. You are paying for the exploitation of the planet, you are paying for child labour, you are paying for toxic dyes & synthetic fabrics—all things that are significant costs to our planet, our society and your own health too. Once you factor all that in, picking sustainable fashion is a no-brainer. Choosing conscious fashion will reduce, eliminate and eventually reverse this impact. By picking natural fibres and local supply chains, paying fair wages and stopping all forms of exploitation and modern slavery, we'll slowly but steadily fix this problem. Brands that are not sustainable will fall behind in the market and will need to adapt and adopt sustainable practices to cater to their customer's demands.'

Preloved and Preowned: Thrifting

Looking good on a budget doesn't always have to mean hurting our planet. Thrifting ultimately lightens both the weight on your pocket as well as on our precious natural ecosystem.

Thrift stores function on donations and re-sell used clothing at low prices. This prevents clothes that would have once ended up in landfills from causing harm to the planet. Collectively, thrifters are helping to divert 3,00,000 tonnes of textiles from landfills each year.[52]

According to Goodwill, the average cost of a woman's blazer at one of its stores is just $4.99[53] versus the price at retailer Express, which can come out to more than $100.[54]

The best part? Thrifting is a lot like treasure hunting—every now and then, you are sure to find a vintage piece that is worth more than its weight in gold and fashion history.

Indian thrifting culture is still in its nascent stages but is quickly picking up the pace. Today, reliable thrift stores, such as Bodements, Ecotopia and Paradime can be found digitally via social media. In India, a number of physical thrift stores and events have opened their doors to the public too, such as

Ciceroni's Preloved Garage Sale and Bombay Closet Cleanse.

Buy Local and Buy Less

If, like me, you are comfortable with living a minimalistic lifestyle, you are already removing yourself from the harmful cycle of pollution and exploitation.

Slow fashion is the sustainable answer to today's use-and-throw mentalities and is built on the principle that the longer an item of clothing takes to become unusable, the better it is for the environment.

In monetary terms, over $500 billion is lost annually due to the underutilization of clothing as well as a lack of recycling.[55]

Minimalistic clothing is simple and requires a one-time investment in a high-quality product that can last longer.

Buying local can further help reduce our carbon footprint due to the lack of pollution associated with importing, and allows us to get an eyewitness account of how the clothes are made, and how people along the supply chain are treated.

Reuse, Donate, Recycle

Clothes can be reused by the age-old practice of familial hand-me-downs and donations to stores or to non-profits, which then pass them on to the marginalized members of society. India has a thriving culture of hand-me-downs, which is considered a rite of passage for siblings and children as they come of age.

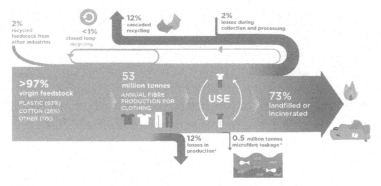

Figure 3-16: Annual fibre production for clothing and the various outcomes.

If the life of one item of clothing could be extended by nine months, we would reduce carbon and water footprints by around 20–30 per cent each.

For thrifters, a marginal 10 per cent increase in sales of second-hand clothes would deliver compounding benefits for our environment, cutting emissions per tonne of clothing by 3 per cent, and lowering water use by 4 per cent.[56]

Try this:

Next time you're clearing out your closet, think of the many ways your clothes can be up or downcycled. Make curtains, rags, tablecloths and quilts from attractive fabrics, and keep them out of landfills.

This sustainable practice doubles as a fun arts-and-crafts project too!

One great example of an organization taking textile-related collateral damage into its own hands is Goonj. With a focus on promoting a circular economy

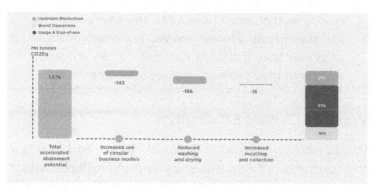

Figure 3-17: Emissions savings within usage and end-of-use under accelerate abatement.[57] *Source: Global Fashion Agenda*

by ensuring the maximum use of each material, Goonj collects fabrics and used textiles from thousands of donors around the nation, and processes and segregates them before donating, upcycling or properly discarding them entirely. Among their many initiatives directed towards building a sustainable future for the underprivileged and economically and socially disparate in India, Goonj has instated several Dropping Centres[58] and Collection Camps[59] to make donating that much easier, and ethical for the receiver.

Grazia editor-in-chief Mehernaaz Dhondy says:

'When things like "Who Made My Clothes" or "Zero Waste" stop being simple buzzwords and are put into practice, consumers will sit up and take notice. Whether it's mindful purchases, recycling, upcycling, thrifting—See, Buy, Love needs to have an added "Keep" to it—but the longevity of a product depends on how well it has been made, produced using resources that aren't harmful to the environment or socially exploiting those involved in the fashion chain of production. If brands can put out an authentic story with their product, we may see some changes.'

Washing Clothes Efficiently

Warm water weakens the fibres of clothes and results in the formation of microplastics and microfibres, which can cause a ripple effect that threatens the sanctity of our water bodies and our health.

On the other hand, two weekly loads of laundry washed in cold water can prevent approximately 500 pounds of CO_2 emissions annually. Wash your clothes in cold water, using cold water detergents that are designed to clean better in cold water.[60]

Wardrobe Detox

Dyed clothes contain soluble chemicals which are carcinogenic and can lead to health problems long after a purchase is made.

Wearing fabrics that aren't dyed prevents these chemicals from being absorbed by our skin and also prevents water wastage from continued production. Wearing naturally dyed clothes instead of conventionally dyed ones prevents 25 to 40 gallons (95 to 150 litres) of water wastage for every 2 pounds (about a kilo) of fabric.[61]

Choose your wardrobe carefully, and do a regular detox of harmful chemicals from your closet. The planet will thank you for it, as will your body.

Disposal Mechanisms

Just like the greenwashing-recycling programme, several textile companies have started their own incentive-based initiatives to get consumers to give back their clothing items, which are then upcycled into new products via textile recycling bins.

While this is effective to a certain extent, to truly avoid greenwashing at the hands of marketing giant retailers, the best way to truly dispose of unused clothing is to pass it on, swap it or donate it, since downcycling weakens the fibres and may cause further deterioration of the cloth.

Change Your Mindset

Social media has contributed significantly to fast fashion, with retailers amping up their processes to meet the ever-growing demand of a global audience that is always online.

While the social media monster cannot be tamed, a change in mindset might help us befriend the beast.

Social media influencers have a responsibility to guide their followers and must amplify slow fashion by illustrating the various benefits of minimalism and promoting DIY clothing trends.

In fact, this trend is already turning on its head, with many brands moving to more sustainable production methods. As of May 2018, 12.5 per cent of the global fashion market had pledged to make sustainable changes by 2020.[62]

So, whether you own four pairs of clothes, forty or 400, one sustainable move today can mean a landslide of good news for the future.

Remember, a stitch in time can save the lives of many.

Resources

Here is an inventory of supplementary media material, people who matter and shops in line with the chapter's theme.

Books

1. *Unraveled: The Life and Death of a Garment*
    Author: Maxine Bedat
    Published by: Portfolio
    https://www.amazon.in/Unraveled-Death-Garment-Maxine-Bedat-ebook/dp/B08JKRM3TH

2. *Overdressed: The Shockingly High Cost of Cheap Fashion*
    Author: Elizabeth L. Cline
    Published by: Portfolio
    https://www.amazon.in/Overdressed-Elizabeth-Cline/dp/1591846544

3. *Clothing Poverty: The Hidden World of Fast Fashion and Second-Hand Clothes*
    Author: Andrew Brooks
    Published by: Zed Books
    https://www.amazon.in/Clothing-Poverty-Fashion-Second-Hand-Clothes/dp/1783600675

4. *The Conscious Closet: The Revolutionary Guide to Looking Good While Doing Good*
    Elizabeth L. Cline

Published by: Plume https://www.amazon.in/
Conscious-Closet-Revolutionary-Guide-Looking/
dp/1524744301

5. *To Die For: Is Fashion Wearing Out the World?*
    Lucy Siegle
    Published by: Fourth Estate
    https://www.amazon.in/Die-Fashion-Wearing-
    Out-World/dp/0007264097

Documentaries

1. *Fast Fashion: The Shady World of Cheap Clothing*
    Directed by: DW Team
    Available on DW Documentary's YouTube Channel
    https://www.youtube.com/watch?v=YhPPP_
    w3kNo

2. *Unravel: The Final Resting Place of Your Cast-Off
    Clothing*
    Directed by: Meghna Gupta
    Available on both Aeon Video and YouTube
    https://aeon.co/videos/this-is-the-final-resting-
    place-of-your-cast-off-clothing

3. *Made in Bangladesh*
    Directed by: Rubaiyat Hossain
    Available on Amazon Prime Video
    https://www.made-in-bangladesh-movie.com

4. *The River Blue: Can Fashion Save the Planet*
   Directed by: Roger William and David Mcllvride
   Available on Amazon Prime Video
   https://riverbluethemovie.eco

5. *Udita Arise*
   Directed by: Hannan Majid and Richard York
   Available on YouTube
   https://www.youtube.com/watch?v=g_tuv
   BHr6WU

TedTalks

1. Three Creative Ways to Fix Fashion's Waste
   problem
   By: Amit Kalra
   https://www.youtube.com/watch?v=yeVU2Ff4ffc

2. I Broke Up With Fast Fashion and You Should Too
   By: Gabriella Smith
   https://www.youtube.com/watch?v=mKPB0
   uW4cto

3. Change Your Closet Change Your Life
   By: Gillian Dunn
   https://www.youtube.com/watch?v=WiVHSRY
   2I5Y

4. The High Cost of Our Cheap Fashion
   By: Maxine Bedat
   https://www.youtube.com/watch?v=5r8V4
   QWwxf0

5. How to Engage With Ethical Fashion
    By: Clara Vuletich
    https://www.youtube.com/watch?v=WXOd4qh3JKk

People

1. Aditi Mayer | @aditimayer
    Talks about sustainability in fashion, social justice,
    and labour rights
    https://www.instagram.com/aditimayer

2. Anya Gupta | @anya.gupta
    Talks about Slow Fashion
    https://www.instagram.com/anya.gupta

3. Elizabeth Joy | @consciousstyle
    Talks about fair fashion and runs a podcast on
    conscious style
    https://www.instagram.com/consciousstyle

4. Alexis | @girlwithgreencloset
    Sustainable Fashion
    https://www.instagram.com/girlwithagreencloset

5. Venetia La Manna | @venetialamanna
    Fair Fashion Campaigner
    https://www.instagram.com/venetialamanna

Shops

1. No Nasties
    For Males and Females | Petite and Plus Sizes

Delivers PAN India
https://www.nonasties.in

2. Doodleage
   For Females and Kids
   Store in Delhi. Online available across India
   https://doodlage.in

3. IKKIVI | Marketplace for sustainable fashion brands from India
   For Males and Females
   Delivers PAN India
   https://www.ikkivi.com

4. Eco Clothing India
   For Females
   Delivers PAN India
   https://ecoclothingindia.com

5. The Summer House
   For Females
   Delivers PAN India
   https://thesummerhouse.in

6. Upcycle luxe
   Sustainable fashion marketplace
   Delivers PAN India
   https://upcycleluxe.com/

7. Earthy Route
   For Males and Females

Delivers PAN India
https://earthyroute.com

8. Nicobar
For Males and Females
Delivers PAN India
https://www.nicobar.com

9. Refash
Upcycled Fashion and Accessories
Delivers PAN India
https://refash.in

10. Malai Eco
Vegan and Compostable Leather Material and Products
Delivers PAN India
https://malai.eco

11. Studio Beej
Sustainable Fashion
Delivers PAN India
https://studiobeej.com

12. Resistor
Sustainable Women Clothing
Delivers Worldwide
https://reistor.com

13. Ecoright
Sustainable Bags, Phone cases, and more

Delivers PAN India
https://www.ecoright.com

14. Urth
Sustainable fashion for Women
Delivers PAN India
https://www.urthlabel.com/

15. Murtle
Upcycled Footwear
Delivers PAN India
https://murtle.in

16. Aulive
Leather alternative products
Delivers PAN India
https://www.aulive.in/

Donate your clothes here

1. Goonj
Multiple cities
https://goonj.org/

2. Clothes Box Foundation
Multiple cities
https://clothesboxfoundation.org/

3. Happiee Souls
Doorstep pickup for old goods
https://happieesouls.com/

Thrift Stores

1. Bombay Closet Cleanse
    sell/swap your clothes
    https://www.instagram.com/bombay
    closetcleanse/

2. Curated Findings
    For both women and men
    https://www.instagram.com/curated.findings/

3. Vintage Laundry
    For both women and men
    https://www.instagram.com/vintage.laundryy/

4. Panda Picked
    For women
    https://www.instagram.com/pandapickedstore/

4

Transportation and Daily Commute

GHG Emissions: 21 per cent[1]

This industry impacts the following Sustainable Development Goals (SDGs):
- SDG 12: Responsible Consumption and Production
- SDG 13: Climate Action

ACTIVITY

1. Where do you do most of your shopping?
    - ☐ Online
    - ☐ Physical Stores
    - ☐ Other

2. How often do you import everyday items from foreign countries?
    - ☐ Sometimes
    - ☐ All the time
    - ☐ Never

3. Do you believe your individual choices make a difference?
    - ☐ Yes
    - ☐ No
    - ☐ Other

4. How often do you take your vehicle for servicing?
    - ☐ Once a Month
    - ☐ Once in Two Months
    - ☐ Once in Three Months
    - ☐ Once in Six Months

5. Do you use any of the following cab/auto services?
    - ☐ Ola/Uber
    - ☐ BluSmart/Evera
    - ☐ Other Rental Cabs
    - ☐ None of the above

Try this:

Take stock of your belongings and answer this question as honestly as you can.

How many of the following do you own?

- A laptop
- A mobile phone
- A car
- A motorcycle
- Edible oil
- Luxury apparel items
- Avocados
- Optical and medical instruments
- Toys
- Fine jewellery

Now, do a quick Google search or read the labels on the products and make a tally of how many were Indian-made, and how many were manufactured elsewhere.

Whether we know it or not, our consumption patterns have a much larger effect than we may realize. Every one of the items on your list that was imported from another country—sometimes even those that are manufactured in India—was likely made of parts that were, in turn, imported from another country. With each import and each export, the overall carbon footprint of your product increases.

This means that by the time your fancy new laptop reaches you from its manufacturing hub, its carbon footprint is roughly the equivalent of a small car driven for approximately 6,140 km.[2]

In 2020, India was number 18 in total exports and number 12 in total imports. Every year, the average Indian household imports a plethora of products, ranging from the luxurious and exotic to the everyday.[3]

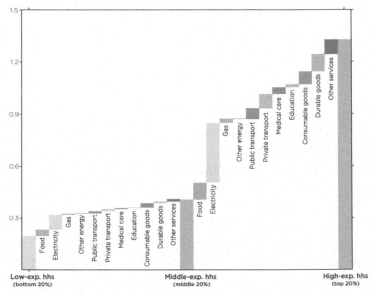

Figure 4-1: Household consumption carbon footprint per capita.[4] *Source: ScienceDirect*

Of the total Indian imports, 21 per cent comprise petroleum and crude oil, 13.4 per cent comprise electronic goods, medicines account for 8.8 per cent, and machinery and chemicals account for 8.1 per cent and 5 per cent, respectively.

Conversely, India's top four exports include engineering goods at 24.6 per cent, gems, jewellery and petroleum products at 8.9 per cent and drugs and pharmaceuticals at 8.4 per cent.[5]

But why am I telling you all of this? Let me explain.

As you read this chapter, you may find yourself struggling with feelings of helplessness, as if the ills of the transportation industry are out of your hands—much like I felt when I was studying it.

But does that mean we truly are helpless? Does our perspective even matter?

If there is one thing I have learnt over my years as an activist, it is that change occurs on many levels.

From governments that make policies to companies that modify their ESG standards to suit ever-changing industry norms, and from global public outcries to natural occurrences that change the face of our lives as we know it—change never, ever occurs in a silo. However, it is important to remember that change, no matter how big, small, isolated or widespread, begins with just one person making a choice.

Demand influences supply, and as consumers, we have more influence than we know.

If you have ever started a business or worked in marketing, you are probably familiar with the phrase 'the customer is king'. The capitalistic structures that we all occupy are built upon this very notion.

From the name of a company to the colours of its logo, and from the products it supplies to its advertising campaigns—the consumer's preferences have the power to build empires and raze them to the ground.

So why does your perspective matter? It matters because the market caters to *you*.

Remember, 7.8 billion people once thought, 'It's only one plastic bottle.'

The modern world is defined by accessibility. From the world wide web, making aeons of human history available to us at just the click of a button, to goods and services delivered to your doorstep in mere minutes after the need for them arises, our quest for accessibility has given rise to a whole new industry—that of commercial transportation.

In 2019, the global transportation services industry group had total revenues of $2,207,827.8 million, representing an annual growth rate of 4 per cent from 2015. According to a 2020 estimate, the transportation and logistics industry employs 450 million people all over the globe.[6]

Today, e-commerce companies and aggregators have the infrastructural capacity to offer twenty-four-hour delivery across the globe, regardless of where the product originated from—and they do it without overburdening their wallets or that of their consumers.

However, there are hidden costs to the convenience we hold in such high regard.

The transportation industry accounts for about 21 per cent of all GHG emissions, releasing several million tons of greenhouse gases each year into the atmosphere.[7]

As a responsible consumer, I have seen the impact of profit-centric capitalism with my own eyes. From the rampant generation of dangerous gasses and emissions to the plastic waste generated by truckfuls of same-day deliveries—I am all too familiar with the collateral damage this never-ending quest for convenience has wrought.

At the moment, I am writing to you from my current home in Leh, Ladakh. I have three meals a day, and all the luxuries I could possibly need. I am an award-winning content creator and artist, and I have the support of a community of over 40,000 people.

And the best part? I did it all with close to zero impact on the environment.

Let me tell you how.

From Here to There and Back Again—The Problems with Transportation

When we think of transportation-related impacts on our environment, we tend to imagine black plumes of smoke rising from exhaust pipes or rampant fossil fuel consumption. However, the impact begins right from the source.

Let's examine some of the less publicized sources of unsustainability that we, as consumers, are never made aware of, as well as a few that we are all too familiar with.

Transport Facilities

Whether it's airports, train stations, warehouses or shipping yards, the infrastructure and maintenance of transport facilities demand the use of vast expanses of land, energy and raw materials to keep things running efficiently. For example, logistics warehouses for products provided by online retailers must be kept running 24/7, making use of round-the-clock electricity, fuel and more. The packaging materials required for the transportation of goods too are typically non-sustainable in nature and are often improperly discarded both at the source as well as at the receiver's end.

Manufacture of Vehicles

From cars and buses to planes and ships, manufacturing large vehicles and heavy machinery both for commercial and personal use is a resource and energy-heavy

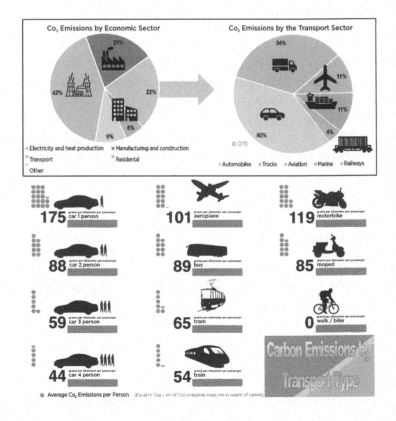

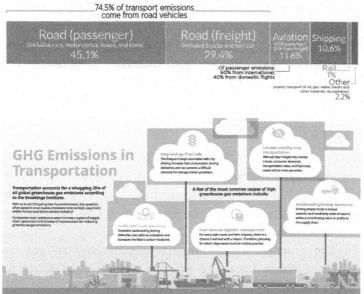

Global Co$_2$ emissions from transport
This is based on global transport emission 2018, which totalled 8 billion tonnes Co$_2$.
Transport accounts for 24% Co$_2$ emissions from energy.

74.5% of transport emissions
come from road vehicles

| Road (passenger) (includes cars, motorcycles, buses, and taxis) 45.1% | Road (freight) (includes trucks and lorries) 29.4% | Aviation (81% passenger, 19% from freight) 11.6% | Shipping 10.6% |

Of passenger emissions:
60% from international;
40% from domestic flights

Rail 1%

Other (mainly transport of oil, gas, water, steam and other materials via pipelines) 2.2%

GHG Emissions in Transportation

Transportation accounts for a whopping 25% of all global greenhouse gas emissions according to the Brookings Institute.

With so much CO2 going into the environment, the question often asked is what causes emissions to be so high, especially within the transportation service industry?

Companies must continue to examine every aspect of supply chain operations to find areas of improvement for reducing greenhouse gas emissions.

A few of the most common causes of high greenhouse gas emissions include:

Stop-and-go final mile
The frequent stops associated with city driving increase fuel consumption during deliveries and can present a difficult obstacle for transportation providers.

Limited visibility into transportation
Without clear insight into market trends, consumer demands, transportation rates, and fuel costs, waste will be more pervasive.

Inefficient route planning
Excessive backtracking during deliveries also adds to emissions and increases the fleet's carbon footprint.

Poor reverse logistics management
For every sale made and item shipped, there is a chance it will end with a return. Therefore, planning for return shipments must be routine practice.

Deadheading/empty backhaul
Driving empty trucks is always wasteful and needlessly adds emissions without contributing value or profits to the supply chain.

Figure 4-2: Carbon dioxide emission for various sectors.[8][9][10][11] Sources: various

undertaking. Often, the tools and infrastructure in the manufacturing process make use of lethal chemicals and fossil fuels, which are improperly discarded as run-off in our waterbodies.

Operating Vehicles

Internal combustion engine (ICE) vehicles, which currently make up more than 99 per cent of vehicles driven across the globe, run on fossil fuels and other types of non-renewable energy, such as coal and natural gas.[12]

Vehicle Maintenance

Waste generated from maintenance of transportation devices, from private vehicles to commercial aeroplanes, in the form of discarded parts, chemicals, lubricants and more, can cause problems for surrounding communities.

Vehicle Disposal

Every vehicle, whether made for road, sea or air, is limited in its lifecycle. When it is replaced by newer technologies or simply becomes outdated, vehicle disposal is another source of solid waste generation and emissions.

The transportation industry is heavily reliant on non-renewable forms of energy, which contribute to rampant greenhouse gas emissions, with the beginning, middle and end of the lifecycle of most vehicles and associated infrastructures contributing to the damage. Some gases, such as nitrogen oxide, are key culprits in the depletion of our planet's stratospheric ozone (O_3) layer, a natural formation that protects the surface of the earth from ultraviolet radiation.

The transportation industry disproportionately affects millions around the globe, polluting the air, the land and the sea more than any other industry. But what does that mean for the health of our planet, and just as importantly, for our own health?

To answer this question, let's take a short trip, shall we?

In the Air

What was the last thing you ordered over the internet? Was it a pair of new shoes or an appliance? Was it your dog's favourite brand of kibble or fancy make-up from a foreign brand?

A hundred years after the first propelled flight ever witnessed by humankind, air transport of cargo is the fastest way to transport goods, and also one of the most harmful to our planet.

Aviation is responsible for approximately 2.4 per cent of global CO_2 emissions. When combined with other gases and water vapour trails produced by aircraft, the aviation industry contributes to about

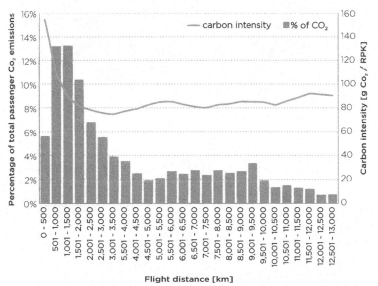

Share of passenger CO_2 emissions and carbon intensity in 2018, by stage length.

Figure 4-3: Share of passenger carbon dioxide emissions and carbon intensity in 2018 by state length.[13] *Source: Our World in Data*

5 per cent of global warming. Although this number seems negligible, it is not just a matter of *how much*, but *how*.[14]

Nitrogen oxides, the emissions most responsible for causing air pollution, are considered harmful on land. When released directly into the upper troposphere at high altitudes, their impact on ozone formation is greatly amplified. Besides this, the water vapour resulting from aviation of any kind released at high altitudes forms ice crystals. These crystals, in turn, catalyse the formation of cirrus clouds—a type of wispy, short and detached formation that is capable

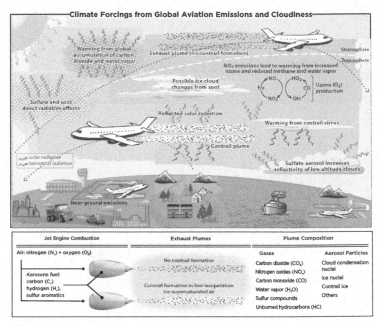

Figure 4-4: Climate forcings from global aviation emissions and cloudiness.[15] *Source: OECD*

of trapping heat—enhancing the greenhouse effect as much as regular GHG emissions.

Air cargo transportation currently accounts for a small portion of global freight, with the number being less than 1 per cent. However, this number is growing exponentially.[16]

Air cargo transportation threatens the sanctity of our environment in three ways:

- Emissions released by an aircraft at take-off and landing contribute to conventional air pollution, as well as global warming

Take-off and Landing

- Emissions released during flight also contribute to global warming

In-Flight

- Noise, material pollution, traffic congestion, and other land-use issues pose serious threats to the natural fabric of areas surrounding airports.

Around Airports

The air transportation industry is disproportionately responsible for our decreasing air quality due to toxic pollutants and particulate matter released into the air. These air pollutants are associated with cancer, cardiovascular, respiratory and neurological diseases.

Let's look at some of these air pollutants and how they affect our wellbeing.

Now that we have travelled the air and seen the effect of pollution from transportation on our air quality, let's embark on the next leg of our adventure: into the ocean!

In the Ocean

Ever wonder where the term 'shipped from' originates? While the term is not necessarily always used in the

Carbon monoxide (CO):

• Originating from cars, trucks and other vehicles or machinery that burn fossil fuels, it is one of the most harmful gasses, which, when inhaled, reduces the availability of oxygen in the circulatory system that sustains critical oceans like the heart and brain.

• At specific concentrations, CO can cause dizziness, confusion, unconsciousness and even death.

Nitrogen dioxide:

• (NO_2) emissions from fuel can reduce overall lung function, affect the respiratory immune defence system and increase the risk of respiratory problems such as asthma.

Sulphur dioxide (So_2) and nitrogen oxides (NO_2):

• Combine in the atmosphere to form various acidic compounds that, when mixed in cloud water, create acid rain.

• Acid precipitation has detrimental effects on our infrastructure and environment, reduces agricultural crop yeilds, and causes forest decline.

Smog:

• Mixture of solid and liquid fog and smog particles formed through the accumulation of carbon monoxide, ozone, hydrocarbons, volatile organic compounds, nitrogen oxides, sulfur oxide, water, particles, and other chemical pollutants.

• Especially severe in cities like Delhi.

• Causes reduction in visibility, negatively impacts air quality.

• Poses health risks such as respiratory problems, skin irritations, eye inflammations, blood clotting, and various types of allergies.

context of marine transportation, once upon a time, this was very much the case.

Today, marine transportation drives a big chunk of global trade, moving approximately 11 billion tonnes of containers of solid and liquid bulk cargo across our oceans every year. This figure represents an annual 1.5 tonnes per person across the global population.[17]

For a long time, maritime operations were largely unregulated from an environmental standpoint. However, all of that changed in the 1960s, when accidental oil spills caused widespread coastal pollution and resulted in the deaths of thousands of seabirds and marine species.[18]

This was the birth of the International Convention for the Prevention of Pollution from Ships (MARPOL), which holds organizations accountable for both accidental and operational marine pollution.

Even with MARPOL in place, however, we have more to be concerned about than oil spills.

For example, in 1992, a shipping container packed with 29,000 yellow rubber ducks was lost at sea on its journey from Hong Kong to the US. This resulted in an all-encompassing flotilla of yellow rubber duckies that soon became known in the media as 'the Friendly Floatees', and are credited with enhancing our understanding of ocean currents—having been found in the faraway lands of Alaska, Indonesia, Hawaii and more.[19]

While this story is an interesting one for scientists and researchers, those rubber duckies also represent a much more troubling issue—twenty years later, they are still washing ashore in near-perfect condition, a jarring metaphor for the pollution caused by rampant consumerism.

This is not to mention other kinds of marine transportation, such as gas and fuel lines running in the depths of the ocean, which famously led to the Gulf of Mexico 'Eye of Fire' incident as early as last year when a gas leak in an underwater pipe resulted in the ocean seemingly catching fire, the flames being witnessed bubbling to the surface of the water.[20]

Marine transportation, by virtue of its frequency and infrastructure, contributes greatly to the damage caused to marine environments.

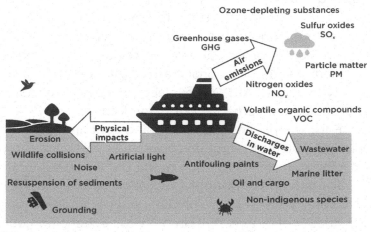

Figure 4-5: Damage caused by marine vehicles.[21] Source: ScienceDirect
Source:

Decreased water quality has become increasingly prevalent due to maritime operations, such as dredging, i.e., the manual deepening of a harbour to facilitate ship movement via the removal of sediments from a riverbed, and accidental oil spills, among a host of other contributors.

Let's examine some of these in further detail:

Emissions

GHG emissions and exhaust gas from ship combustion engines are a major cause for concern, with annual estimated emissions of 1 billion tons of carbon dioxide equivalents (CO_2eq). This contributes to air pollution as well as, in some cases, acidified water.[22]

Oil Spills

Oil spills, both from operational efforts and accidents, can be highly detrimental to marine life and can be notoriously difficult to clean up, with many efforts continuing to this day. Oil can destroy the insulating ability of many furry animal coats, and can hinder the natural water repellent skin of marine birds, exposing both to potentially lethal environmental conditions. Oil from spills is often ingested by animals, like dolphins and birds, who suffer from lowered immune, reproductive and lung functions, as well as poisoning. Fish populations have been observed to experience a number of adverse effects, from reduced levels of growth to changes in heart and respiration rates, as well as more apparent physical changes, such as fin erosion, enlarged livers and impaired reproduction. Fish eggs and larvae are sensitive to even the most minor changes in pH levels, and when exposed to toxic chemicals and oils, can experience mutations or even die.

Underwater Noise Pollution

While most are aware of noise pollution in big cities, many remain ignorant of the impact noise pollution can have on our marine ecosystems. Ocean noise pollution results from human activities, like seismic surveys, oil exploration, military sonar and most prevalently, commercial shipping. Oceanic noise pollution can result in a range of serious threats to marine life, dramatically altering behaviour due to stress. It can reduce an animal's ability to communicate, navigate, locate prey, avoid predators and find mates, and can even result in creatures being driven from their natural habitats, putting them at mortal risk from carnivorous species.

From the air to the ocean, transportation can have harmful, and sometimes deadly, consequences for our environment.

Did any of these facts surprise you? As the saying goes, out of sight, out of mind—but what about what we can see, and what plagues us every day?

This brings us to our next section: on ground.

On Ground

Land transportation comes in many forms, and most of them come with their own negative impacts on our environment.

When we think of pollution on land, the first thing that comes to mind is emissions from private vehicles—and for good reason. A standard passenger vehicle emits about 4.6 metric tons of carbon dioxide per year via the burning of fossil fuels, such as petrol and diesel.[23]

In addition to this, private automobiles are also responsible for other dangerous emissions, such as methane, nitrous oxide and hydrofluorocarbons. While these are smaller in quantities than CO_2 emissions, they have a higher global warming potential, making them unsustainable even in small quantities.

Commercial land transport usually occurs by truck or rail. Trucking has garnered the attention of environmentalists from around the world for its unsustainability on two fronts—air pollution and noise pollution. The global road freight sector is responsible for 9 per cent of global greenhouse gas emissions.[24] According to a study by WEF and McKinsey, road freight now accounts for 53 per cent of CO_2 emissions within global trade-related transport, and this share is expected to rise to 56 per cent by 2050 if current trends continue.[25]

Alkem Laboratories managing director Sandeep Singh says:

'While everyone is focused on building capacity and infrastructure, it is the basic science that needs to be revisited. Prioritizing green chemistry in engineering, OEMs must have a laser-sharp focus on eliminating the use and generation of hazardous substances in the initial design of the product itself. India is the world's pharmacy, and as such, we cater to a large population both nationally, as well as internationally. As one of the largest pharmaceutical markets in the world, and as one of the main suppliers to the globe, investing in sustainably produced generic medicine presents a powerful economic opportunity.'

On the other hand, rail transport is considered to be more environmentally friendly, but still has negative effects on the surrounding areas, which suffer from noise pollution-related concerns.

However, what most people tend to miss when speaking of land pollution is not the cars, trucks, heavy vehicles, or even trains—but rather what is *beneath* them.

Road and highway construction, as well as other development activities, can disrupt the natural balance of surrounding areas in potentially life-threatening ways.

Let's examine a few of these now.

Biodiversity

Large-scale development projects aim to reduce travel time for private and commercial vehicles, i.e., cutting through the shortest route possible. More often than not, that route cuts through delicate ecosystems hitherto untouched by human interference. When highways and roads cut through forested and fertile land, they not only lead to mass deforestation, but also damage the natural landscape, habitat and biodiversity of the area, and irreversibly change the conditions of the area itself by increasing noise pollution, underground vibrations and more.

Culture and Society

Large-scale development projects are also responsible for the displacement of indigenous communities, who tend to have a symbiotic relationship with the land—caring for it as much as it cares for them. The destruction of the cultural and social structure of affected communities in this way is irreversible and leads to a loss of sustainable and symbiotic natural and human relationships.

Noise

The noise and vibration generated from the thoroughfare of cars, vehicles, trains and more may not seem like much to the average person, but can greatly disrupt the environment around it, hindering the movement of animals, confusing their natural threat detection systems and even driving them from their natural habitats, threatening not only the populations of wild flora and fauna but also surrounding human settlements due to increased levels of human-animal conflict.

Soil, Air and Water Quality

Microparticles generated by the maintenance and use of roads, pavements and other infrastructural endeavours enter the soil with the leaching of water, compromising its integrity. Pollutants and particulate matter resulting from vehicle exhausts and more are often transferred by air and dispersed over varying distances, and can also contaminate the soil. Fuel emissions and oil leaks associated with motor vehicles, as well as chemicals used for the preservation of rails, also make their way into the soil, being non-soluble and carried with water, leading to contamination. The construction activities, noise and usage-associated vibration may also enhance soil erosion, depleting its natural fertility and percolative abilities.

What Are We Missing?

In today's hyper-connected world, it can be hard to zoom out and assess how our choices affect others on both ends of the supply and consumption chain. We are so focused on immediate gratification that we often forget what it takes to make it all happen.

However, we must also realize that the industry is *designed* to make us not think about what goes on behind the scenes.

In the modern world, material belongings are heralded as a sign of luxury and status—the more you have, the more valuable you are as a member of society. However, this is far from the case.

As I look around my room now, I can see my whole life at a glance. I mean this literally. I can count on one hand the number of clothes I own, including my shoes. I own no fancy goods or imported keepsakes. In terms of their carbon footprints, my most taxing belongings are my electronics—an Apple phone and laptop, and my Sony camera and shooting equipment. For each of these, I have offset their carbon footprint by various means, including, but not limited to, donating to carbon offsetting charities, reducing my carbon footprint and planting trees.

I live a highly minimalistic lifestyle and always have. That is not to say I am better than someone who likes material objects—because I am not.

When I was younger, I longed for many material possessions. I wanted a bike, a pair of light-up sneakers, fancy clothes and much more. However, I came from a modest background and knew that money must be spent frugally.

As I grew older, I realized I didn't need as much as I thought I did. As I learnt more about sustainability, I stopped wanting the things I knew I didn't need.

When giant multinational corporations with employee bases as large as the population size of a small country do not bat an eyelid, when oil and ICE vehicle manufacturers shirk responsibility for devastating environmental accidents, and fight the shift toward more sustainable options for their products, how can we, as individuals, even attempt to fix the problem?

Well, my answer is this: their lack of accountability does not reflect your sense of responsibility. We are all accountable for our actions, and that is the only thing we need to know in the quest to make the world a safer and more sustainable place for future generations.

Climate change is a global problem, and no one nation, no individual can be faulted for it. In much the same way, our consumption habits contribute to the issue on a global scale. While I cannot ask you to bear the weight of reforming the industry all on your own, I must remind you that collective change is required to fix a collective problem—and change, always, begins with just one person.

As I look around my room, I can tell you that I love my things. However, I am not attached to them. As a nomad, I have made many homes and many lives. However, when you make a new home for yourself, you can only carry so much with you. With every home I have made for myself, there is another left behind, and with it, all the things I could not carry with my two hands. This is the core of minimalism, to me—to keep only what you need, and to give the rest away to someone who needs it more than you do.

While a minimalistic lifestyle is a bizarre concept in today's globalized market, there is a less extreme way to combat the adverse effects of commercial transportation on the air, the ocean, the land and all living things.

Global Citizens, Local Lives: Solutions

While it is true that the world has gotten smaller, and we are connected beyond borders and across oceans and mountain ranges, we are still very much a part of our land. As such, we can make big changes by returning to the law of the land, and existing within its boundaries and limits.

Here are a few ways we, as individuals, can reduce the impact commercial transport has on our environment.

Shop Local

Fun Fact: Did you know that these beloved brands and companies are not Indian?

Bata: Undoubtedly one of India's favourite footwear brands, Bata is a symbol of Indian culture—but it's not an Indian brand. Headquartered in Lausanne, Switzerland, the company was founded in 1894 in Zlín, Moravia by a family of cobblers.

Hindustan Unilever: Hindustan Unilever is a respected manufacturer of soaps, detergents and many more household goods. While the title of Hindustan is indeed indicative of their Indian leg, the company is a subsidiary of Unilever, which is a British-Dutch conglomerate.

Colgate: Synonymous with toothpaste, this dental hygiene brand is owned and manufactured by Colgate-Palmolive, an American worldwide consumer products company.

Bose Speakers: Founded by and named after Amar Bose, an American of Indian descent, back in 1964, this world-renowned audio products company is an American brand based in Framingham, Massachusetts.

Maggi: Headquartered in Vevey, Vaud, Switzerland, everyone's favourite two-minute noodle brand is owned by Nestlé, a Swiss multinational company.

Lifebuoy/Tide: Owned by Procter and Gamble, better known as P&G, Lifebuoy finds its origins in England, where it came into existence in 1895.

What separates these companies from their competitors is that they adapted to the Indian market so well that we forgot that they are, in fact, foreign entities with an operating base in India. These companies may not necessarily be sustainable but are great models of listening to the needs and desires of consumers—living, breathing examples of how our choices can change the world.

While it is indeed enticing to own exotic fabrics and materials imported from distant lands, there is more to these seemingly innocent indulgences than meets the eye.

Wherever possible, buy local and seasonal. This will not only cut down on your carbon footprint, but will support local businesses, boost your national economy, and more often than not, result in higher quality, longer-lasting products.

The Indian government's Make In India initiative goes a long way towards reducing both costs as well as emissions by encouraging manufacturers to innovate, ideate and produce locally, making it simpler for you, as a consumer, to make more conscious choices. As recently reported, the import of toys in India is down by 70 per cent over the last three years.[26]

Now, let's take a look at some brands that sound foreign, but are home-grown examples of premium marketing efforts that build and reinforce a narrative of luxury:

Monte Carlo: Established in 1984, this leading clothing brand is known for its woollen apparel. The Italian brand name was specifically chosen to sound alluring to the Indian consumer, keeping in mind their fascination with foreign brands. However, the brand is 100 per cent Indian in origin and is operated by the Ludhiana-based Nahar Group.

Louis Philippe: This premier men's apparel brand name is inspired by French king Louis Philippe, but has its origins on Indian shores, having launched in the country in 1989 and owned by Madura Fashion and Lifestyle.

Amrut Single Malt: Widely regarded as one of the best single malt whiskies available the world over, this fine liquor is brewed and bottled at the Amrut distillery in Bengaluru.

Lakmé: A leader in Indian cosmetics, Lakmé is a 100 per cent subsidiary of Tata Oil Mills and was named after the French opera *Lakmé*.

La Opala: Pioneering high-end tableware in India, La Opala's French-sounding name adds to its brand image. However, all their products, as well as the company itself, are Indian, with manufacturing and headquarters in Kolkata.

Allen Solly: Often mistaken as an international brand due to its name, Allen Solly is a subsidiary of the Aditya Birla Group and is licensed under Madura Garments.

What these brands demonstrate is that luxury and quality are not synonymous with foreign shores. We are a nation of fine craftsmen and artisans who have been practising their craft for generations upon generations. As conscious citizens of our nation, it is our duty to endorse and uplift our countrymen. Do your research, and try to find local alternatives to the products you seek.

Local Transportation

Besides commercial transportation, private vehicles are the biggest source of transportation-related environmental impacts. From daily commutes to work and back, to congested roads and traffic in major metropolitan cities, greening your daily commute can go a long way toward limiting your personal transportation footprint.

For longer distances, EVs and electric two-wheelers are an eco-friendly choice, helping reduce your carbon footprint by 40-50 per cent[27]. While the Indian market is still in its nascent stages, the demand for two-wheelers has skyrocketed in recent times. With around 21 million sales in 2018-19 alone, vehicle start-ups are now introducing their electric versions in the market.[28]

One such start-up is Yulu, a technology-driven mobility platform that enables Integrated Urban Mobility across public and private modes of transport.

Using Micro Mobility Vehicles (MMVs) through a user-friendly mobile app, it aims to tackle challenges of urban mobility, chaotic congestion and sustainable living.

According to Santosh Iyer, managing director and chief executive officer of Mercedes-Benz India:

> 'At Mercedes-Benz, we believe that there is no luxury without sustainability. India is taking small but significant steps towards the transition to EV, and the government's focus on policy interventions in terms of reduction of GST and waiving off-road tax is a step in the right direction. At Mercedes Benz, we have defined a roadmap for a carbon-neutral fleet in our Global Motto of Ambition 2039—a promise we are working towards with great determination, having already achieved Carbon Neutral Production across our manufacturing plants.'

India's largest smart electric mobility platform, **eBikeGo** is now offering an electronic mode of transportation that aims at striking a balance between the fast-paced competitive life of young Indians and taking care of the environment.

Others in the industry, like **VOGO, OLA and Rapido**, are also attempting to switch to sustainable EVs over fuel-driven vehicles.

Ramesh Somani, editor-in-chief, *TopGear India*:

> 'When people ask me why we should make the switch to EV, my answer is simple—if you want December

to remain cold and May to remain hot, then the only way forward is to switch to EV. Almost every major manufacturer has committed to only produce EVs by 2030. This is a big, bold step and shows the seriousness of auto manufacturers toward reducing their carbon footprint. However, it's not up to just them. To promote the adoption of EVs, governments need to participate to help the consumer with their transition. For example, in Bhutan, EV charging stations are free, making the experience more seamless for new users.'

Public Transportation

To make your commute more environmentally friendly, opt for public transportation.

India is well-known for its public transportation, which is an integral part of the functioning of its economy. In Mumbai alone, the railway system caters to over 7.5 million people daily, while around 1.3 million passengers used buses every day in 2021.[29]

Wherever possible, make use of public transportation to limit your emissions. This is also a great way to experience local life!

Mahesh Babu, chief executive officer, SWITCH Mobility, says:

'Our current public transportation capacity is very poor—take buses for example, we have 1.2 buses per 1000 people, the ideal number is closer to 2.5-3 or above. It's not just a matter of how our public transportation runs, but how many public vehicles are available. Electric double-decker buses can take about 86% more passengers, and the energy consumed is 36% lesser than a regular EV bus. 200 electric double-deckers can remove over 10,000 cars driving to and from work every day, at only a fraction of the cost to the environment.'

Reduce Your Driving Footprint

If public transport is not a possibility, there are numerous ways you can reduce your driving footprint while travelling within your city. For example, carpooling can greatly mitigate air pollution from exhausts, as well as congestion and noise pollution, by reducing the per person carbon footprint of each passenger.

Here are a few more tips on making your private commute more sustainable:

- **Go easy on the gas and brakes:** Driving efficiently conserves fuel, thereby saving you money as well as lowering emissions.
- **Servicing:** Regularly servicing your car will improve its efficiency and per mile average.
- **Check your tires:** Properly inflated tires can have a momentous positive environmental impact. Improving fuel efficiency means less carbon dioxide released into the atmosphere. As Barack Obama once said, 'If everybody kept their tires inflated, that would have a big dent; it would produce as much oil savings as we might be pumping in some of these offshore sites by drilling.'

- **Keep your cool:** Cut down on intensive inter-city driving and reduce your use of air conditioning.
- **Cruise:** Enabling cruise control on long drives is fuel efficient and better for the environment.
- **Keep it light:** Keeping your car light and not overburdened by excess luggage also puts less pressure on energy use, leading to lower emissions.

Offsetting Your Carbon Footprint

Sometimes, unsustainable choices are not a choices but necessities. If you cannot shop locally to reduce your carbon footprint, why not offset it?

There are numerous ways to offset your carbon footprint, from making ten sustainable choices for every unsustainable choice to investing in environmentally friendly technologies or donating to environmental justice groups.

In India, many companies are taking responsibility for the carbon footprint associated with the transportation and delivery of their goods and services. One such company making a conscious change is Zomato, which has pledged to offset its carbon footprint and reduce its emissions through the use of electric scooters, whenever feasible, for delivery.

Amol Jaggi, founder of Blu Smart, says:

'Greening public transport in India, or transitioning to low-emission and zero-emission vehicles in the country's public transportation system can have a significant impact on transportation-related emissions. As a nation of 140.76 crore people, there are several options available to provide mobility for the world's largest population. While global players have entered the market with many private and public transportation solutions, a homegrown, EV-focused solution to transportation can be a game changer for both our national economy, as well as our carbon footprint.'

FlyGreen, a booking agency offering a choice of aviation solutions for travellers, is another example, with a promise to offset the client's carbon emission for free by investing in solar panel projects in India.[30]

Adopt a Minimalistic Worldview

If you, like me, are the sort of person that can do without material possessions, you are already doing your part for the environment. However, minimalism doesn't have to be extreme. The ethos of a minimalistic lifestyle is to keep only what you need and to give away everything else.

Do not replace items because there is a newer version in the market, but only do so when the item you own is no longer usable.

To me, minimalism is a way of life. I keep what I need, and nothing else.

As humans, we are capable of more harm and more good than we could possibly imagine having the capacity for. Each one of us has the power to live within our local ecosystems, while still participating as responsible citizens of the world. We all long to experience all that the world has to offer, taste the flavours of exotic lands and live as if we were geographically agnostic. However, this ambition is just another metaphor for human greed, and we are biting off more than we can chew every day.

People always ask me, what can I alone change about an industry that I could not even comprehend in terms of sheer scale?

To them I say this—if you stole one rupee from someone's account every day, that would amount to 365 rupees in a year. If you stole one rupee from every single person in the world for just one day, you could have 7.8 billion rupees. If you stole that amount every day for an entire year, you would be one of the richest people in the world.

The best part? You would do it without anyone ever noticing.

This is the power of one individual, making one choice, every day.

Remember, change doesn't always have to be a wave. Sometimes, it occurs drop by drop, chiselling away at the foundation of solid rock, until it brings down an empire.

Resources

Here is an inventory of supplementary media material, people who matter and vehicles in line with the chapter's theme.

Documentaries

1. *Death By Death*
    Directed by: Arre Indian-Express
    Available on YouTube and Arre
    https://www.youtube.com/watch?v=WSaz__Hfff
    A&t=786s

2. *Most Polluted Major City on Earth*
    Directed by: ABB Formula E
    Available on YouTube
    https://www.youtube.com/watch?v=mNZI
    dHhdQs8

3. *Evolve to Electric with Tata Motors: The Documentary*
    Directed by: Tata Passenger Electric Mobility Limited
    Available on YouTube
    https://www.youtube.com/watch?v=NiCNI3b7MaE

TedTalks

1. A Carbon-Free Future Starts With Driving Less
    By: Wayne Ting
    https://www.youtube.com/watch?v=7INMrxpc7nw

2. How Cities Are Detoxing Transportation
   By: Monica Arya
   https://www.youtube.com/watch?v=x1Efv_wF5LE&t=134s

3. How Green Hydrogen Could End the Fossil Fuel Era
   By: Vaitea Cowan
   https://www.youtube.com/watch?v=9OLxBvLvCoM

People

1. PlugIn India
   YouTube Channel on All Things EV
   https://www.youtube.com/c/pluginIndia

EV

Cab/Bike Service

1. Blusmart | Cab Service
   Gurugram, South Delhi, Few Areas of Bangalore
   https://blu-smart.com

2. Evera | Taxi/Cab
   Delhi NCR
   https://www.everacabs.com

3. Yulu | Single-Seater Two-Wheeler
   Delhi, Gurugram, Bangalore, Mumbai, Pune, Bhubaneswar
   https://www.yulu.bike

Electric Four-Wheelers in India

1. The EQC
    By Mercedes-EQ
    https://www.mercedes-benz.co.in/passengercars.html?group=all&subgroup=all.BODYTYPE.offroader&view=BODYTYPE

2. Tata Nexon EV
    By Tata Motors
    https://nexonev.tatamotors.com

3. Tiago EV
    By Tata Motors
    https://tiagoev.tatamotors.com

4. Mahindra XUV400 , EV8, EV9
    By Mahindra Electric Automobile
    https://mahindraelectricautomobile.com

5. Jaguar I-Pace
    By Jaguar
    https://www.jaguar.in/jaguar-range/i-pace/index.html

6. Kia EV6
    By Kia
    https://www.kia.com/in/our-vehicles/sonet-ae/showroom.html

7. BMW iX and i4
    By BMW

https://www.kia.com/in/our-vehicles/sonet-ae/
showroom.html

8. MG ZS EV
    By MG Motors
    https://www.mgmotor.co.in/vehicles/mgzsev-
    electric-car-in-india

Electric Two-Wheelers in India

1. Ampere
    E-Scooters
    https://amperevehicles.com

2. Tork Motors
    Electric Motorcycle
    https://booking.torkmotors.com/

3. Vida by Hero
    Electric Scooter
    https://www.vidaworld.com/

4. Ather Energy
    E-Scooters
    https://www.atherenergy.com

5. Pure EV
    E-Scooters and Bikes
    https://pureev.in

6. Ultraviolette
    Electric Motorcycle

https://www.ultraviolette.com

7. Revolt Motors
    AI-Enabled Motorcycle
    https://www.revoltmotors.com

8. VAAN Moto
    Electric Bicycles and Mobility Solutions
    https://vaanmoto.com

EV Charging

1. E-Amrit
    Govt. of India's Portal on EVs | Charging Station Locator
    https://e-amrit.niti.gov.in/home

2. Battery Smart
    Battery Swapping Station for Two and Three-Wheeler EVs
    https://www.batterysmart.in

3. Statiq
    App-Based EV charger Booking
    https://www.statiq.in

4. Exponent Energy
    Zero to Full EV Charging In 15 Minutes
    https://www.exponent.energy

5

Tourism

GHG Impact: 8 per cent[1]

This industry impacts the following Sustainable Development Goals (SDGs):
- SDG 6: Clean Water and Sanitization
- SDG 11: Sustainable Cities and Communities
- SDG 12: Responsible Consumption and Production
- SDG 13: Climate Action
- SDG 15: Life on Land

ACTIVITY

Before we begin this chapter, could you please respond to these questions:

1. How many flights have you taken in the last five years?
    - ☐ 1-5
    - ☐ 5-10
    - ☐ 10-15
    - ☐ Other (Please specify)

2. Do you prefer to travel nationally or internationally?
    - ☐ Local
    - ☐ International

3. Where do you stay/prefer to stay while travelling?
    - ☐ Hotel
    - ☐ Resort
    - ☐ Local Homestay

4. On average, how many hours a day do you travel?
    - ☐ 1-2 hours
    - ☐ 2-3 hours
    - ☐ 3-4 hours
    - ☐ Other (Please specify)

5. What is the most exotic land you have ever travelled to?
    - ☐ _____

Once, man would only travel as far as his two feet could take him. In times of hunger or thirst, his feet carried him farther in search of a kill, a harvest or a stream to drink from. As time slowly inched forward, man created the wheel, which could take more steps than he without tiring as easily. Eventually, that wheel became a horse-drawn carriage, which transformed into a horse-powered vehicle, and eventually, a steam-engine-powered train.

When man had tamed the land, he looked to the seas. Ships carried emissaries of empires across vast oceans. Columbus discovered the Americas in search of India, and De Gama discovered India in search of silk, spices and slaves for the Western world.

As technologies evolved, we looked to the skies. The Wright brothers' flying machine paved the way for commercial aeroplanes, while America and the USSR competed to claim first rights to the vast expanses of space.

Today, travel is more accessible than it has ever been in human history. Where once crossing the ocean was a knightly endeavour only undertaken by the noblest of men, today, one need only swipe a piece of plastic to traverse the globe and witness the farthest reaches of our planet.

As one of the biggest contributors to the global GDP, the tourism industry directly employs nearly 77 million people worldwide, which comprises about 3 per cent of the world's total employment.[2] Globally, travel and tourism's direct contribution to GDP was approximately 5.8 trillion U.S. dollars in 2021.[3] The industry's carbon footprint in India alone is estimated to be over 250 million tons of CO_2 equivalent per year (10 per cent of India's total). This figure is the fourth largest in the world.[4]

Social media is a haven for travellers, who share awe-inspiring and aspirational stories of their travels to exotic lands, many still relatively untouched by human activity. For so many of us, travel can be a cathartic undertaking—a way to escape from the mundanity of the everyday and to broaden our horizons while chasing the next one.

However, there is a darker side to these gram-worthy escapades.

As a responsible traveller and activist for climate change, I have seen the impact of gratuitous tourism with my own eyes. From the rampant generation of dangerous gases to the plastic waste generated by seasonal tourists who come in search of natural beauty, only to leave behind destruction in their wake—I am no stranger to the duality of human indifference.

I have been fortunate enough to travel to the farthest reaches of my country, and have seen beauty in its purest form with my naked eye.

And the best part? I did it with close to zero impact on the environment.

Let me tell you how.

*

I was born in the concrete jungles of the city of Indore, where I spent much of my adolescence. My parents had separated when I was only a child, and my mother and I were perfectly happy living in our little bubble. As a single-parent child from a middle-class family, my first encounter with the concept of travelling was much the same as it was for others who grew up the way I did.

Before long car journeys, flights or even cruises, the Indian Railway afforded me my first glimpse of travel. For a long time, the train was the only form of travel I was even aware of.

I always loved train rides and consider them to be an integral part of our collective childhood as a generation born and raised in India in the nineties. There is a certain cultural heritage to the Indian Railway that is lacking in other countries, and that recalls only the happiest memories of my childhood.

When you first embark on a train journey, the people seated on the bunks around you are strangers. However, all this changes the moment someone pulls out their parcel of packed home food. From *aloo-sabzi* to parathas, *jeera aloo, bhindi masala, theplas* and more, the compartment soon becomes a potluck of home-cooked meals and flavours. Before you know it, the miasma of human sweat and perfume in the compartment is replaced with a medley of aromas of authentic Indian cuisine, and those you once considered strangers transform into friends and companions on the long journey ahead.

Once the feasting and merrymaking are over and the journey settles into its forward momentum, the compartment quiets as people of all ages sit with their eyes glued to the windows, watching the world go by outside. From quaint villages to vast expanses of fields, the view from a train window is, in my eyes, only second to that of travel by foot.

Even as an adult, long train journeys bring back to me a certain childlike sense of wonder. Games of *antakshari* and I Spy are punctuated by delays and unplanned stops owing to maintenance on the tracks or some other mysterious happenings. The journey is an adventure in and of itself.

When I was a little older, I travelled with my mother to visit my estranged father at his home in Haridwar, where he operated a small tours-and-travels service.

Growing up in Indore was like growing up in greyscale. There were buildings and roads and not much else in between. Haridwar, on the other hand, was like a visceral rainbow to the senses.

With a cultural heritage that spans centuries, Haridwar is home to some of the oldest temples and structures in India. It is also a haven of natural beauty, with dense forests, cascading waterfalls and teeming rivers galore.

It was here that I had my first ever encounter with the wild—on a trek with my father amid the greenery of Haridwar. It was there that I truly experienced the beauty and power of nature.

Over the following months, my father took me on many adventures around the nation. I visited the medieval Mughal architectural marvels of Agra and Aligarh, undertook the Hindu pilgrimages of Mathura and Vrindavan and experienced the natural allure of Haridwar.

Soon after, however, my parents stopped speaking—for good this time—and my adventures travelling around India seemed to have been stopped in their tracks.

That was, until I took the initiative myself and took a trip from Indore to Mumbai with nothing but my trusty backpack and a bicycle to keep me company. This was one of the first times I was truly alone on my travels and was one of the most liberating experiences of my life—so much so that I soon embarked on another trip from Indore to Bhutan, the farthest I have ever travelled on a bicycle.

As people began taking notice of my work and adventures, I was invited to travel on their dime—from flights to South India to long-distance train rides, and from countless car journeys to bumpy bus rides to alien lands, I soon had the world at my feet. But it wasn't long before I was met with a startling realization that would cause a momentous shift in my mindset and the way I would travel for the rest of my life.

It all came to a head when I first visited Ladakh in 2015. It was my first time witnessing a topography as unique as this one, where grassy mountains were replaced by towering, barren, sandy dunes, and where frozen glaciers twinkled in the sunlight upon the horizon, in stark contrast to the desert-like expanses that stretched out for miles before them.

By this point, I had travelled far and wide and had gained a reputation for being a digital nomad and creator. I was also an avid enthusiast of sustainability and a self-proclaimed student of climate science. Over my travels, I had grown accustomed to the usual trappings of unsustainable tourism and had even contributed to them myself. From littered streets to plastic waste, black smog from exhaust emissions, and, according to me, 'rude and uncivilized' people who believed they owned the place for having bought a seat on a plane, these unsustainable acts were par for the course as far as I saw it.

This was no different, but having seen it all before, I couldn't have predicted the impact Ladakh would have on me.

As I made my way from village to village, the plastic waste and litter felt like a hard, unmovable stain on the otherwise pristine canvas of the landscape. The pure white snow forming on the sides of the road was stained black with exhaust fumes. Upon the horizon, the glaciers stood like monoliths of white, now stained a brownish-yellow that made my heart sink.

It was like the experience of hearing an uncouth statement or a cuss word from the lips of a pious priest—jarring at best, and deeply unsettling at worst.

It was there that I first truly noticed the impact of unsustainable tourism on our natural world. From that moment, I devoted myself to zero-emission travel and pledged to change my ways in the hope that it would inspire others to do the same.

Let me show you what I have learnt, and how you too can see the world without causing it harm.

Oh! The Places You'll Go—The Problems with the Tourism Industry

Wanderlust is a term that skyrocketed in popularity with the advent of social media. Put simply, the word means 'A strong desire to travel'.

However, travel for leisure is in and of itself a selfish undertaking, which perpetrates a great deal of collateral damage on the locations themselves as well as those who inhabit them. Tourism combines the negative impacts of consumption and disposal of every kind—from water to energy, food, transportation-related emissions and more.

Carbon Footprint of Global Tourism

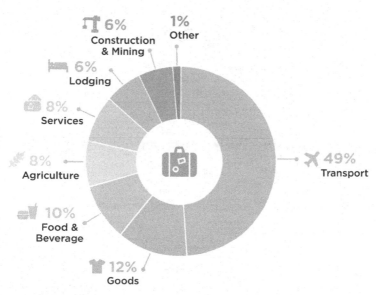

Figure 5-1: Carbon footprint of global tourism. Source: Sustainable Travel International[5]

While many regions such as Andaman and Nicobar, Sri Lanka, Maldives, Jamaica, the Philippines and more are heavily reliant on tourism for their national GDP, the industry itself does as much harm as it does good

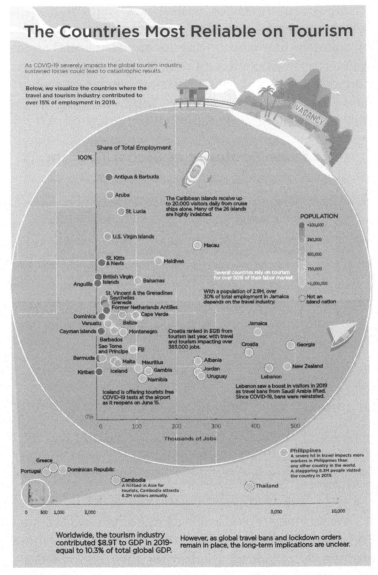

Figure 5-2: The countries most reliant on tourism.[6] Source: Visual Capitalist

to both the locations and the communities that are indigenous to them.

While there are benefits to being known as a tourist destination, the negative impacts associated with the influx of tourists pose a threat to the area's natural ecosystems, making them vulnerable to increased levels of pollution, soil erosion, natural habitat loss and much more.

Let's examine some of these threats through the lens of those affected by them.

Large-Scale Development Projects

Globally, governments allocate massive budgets to development projects in the hope of attracting tourists and their foreign currency.

In our own country, the government recently allocated '2,700 crores (US$422 million) towards building the world's tallest statue, with a height of 182 metres (597 feet). However, the true cost of this project, as well as most large-scale projects of this scale, is much higher.

The statue, for example, has been condemned for its lack of environmental oversight and its displacement of local indigenous people. The land on which the statue was built was once an Adivasi sacred site that was taken from them.[7]

Tourist facilities increase the number of impervious surfaces in an area due to the laying of concrete and tar. This causes more run-off of organic matter, suspended particles, oil and gas, which trickles down into sensitive water bodies and soil, polluting them until they transform into breeding sites for insects and harm marine flora and fauna. Not to mention, these projects prioritize development over preservation, displacing local communities.

The development of supporting infrastructure, such as airports, marinas and more, can also damage wetlands, mangroves, coral reefs and estuaries, which are extremely important for buffering the impact of pollutants in water bodies, which serve as a home to diverse fauna and which protect coastal locations from flooding and other natural disasters.

Wastewater production from building activities and tourists themselves also adds to the stress on sewage treatment plants. In such cases, sewage effluent is directly dumped into neighbouring water bodies leading to eutrophication, i.e., algal growth that may compete with marine flora and fauna for oxygen, sunlight and critical nutrients and minerals.

Gentrification of Cultures and Communities

Unregulated tourism also impacts local communities by putting undue strain on critical resources, such as water, forcing local populations to compete to avail of their rights to them. The much sought after 'local colour' that tourists experience in the form of handicrafts, native food items, rituals and traditions become gentrified, with their cultural and communal values tarnished with a performative dimension, dispelling respect for traditional symbols and sacred practices as a form of entertainment for tourists.

According to several reports, tourism also leads to a boost in human trafficking, with an increase in the number of tourists visiting to engage in sex trade in countries such as Indonesia and Africa. Often, tribes and indigenous groups are exploited for entertainment, with a famous example of this being the Jarawa tribe in A&N, who were made to dance by insensitive tourists in exchange for food.[8]

These indigenous communities have deep roots in these lands, having inhabited and cared for them since the age of their earliest ancestors. Unlike city folk, these communities live in harmony with nature, taking from it only what they need for their survival, and serving and protecting the region as well as its biodiversity from threats. As such, their uncompromised existence in these lands is as essential to the survival and preservation of these regions as any government-mandated conservation laws.

Unsustainable Consumption Patterns

The consumption patterns of seasonal tourists also adversely affect the lands they visit. From packaging and solid waste generated from water bottles to the packets of snacks and other instant food items, and increased energy and water consumption, the tourism industry has a variety of hidden environmental costs.

Here's a homegrown example of a hotspot for tourists. As we can see from the table below, waste generation increases more than ten-fold during weekends, the preferred time for tourists to visit the region.

ARUKAVALLEY, VISHAKHAPATNAM DIST

SOLD WASTE GENERATION	EVERYDAY		WEEKEND	
	R.K. Beach	Aruka Valley	R.K. Beach	Aruka Valley
Total Accumulation	150 kg (0.05 kg)	210 kg (0.09 kg)	5200 kg (0.52)	4600 kg (0.51 kg)
a) Plastic Water Bottles	78.0kg (52%)	78.2 (37.23%)	2704 (52%)	1790 (38.1%)
b) Plastic Covers	30.0kg (20%)	58.8 (28%)	1040 (20%)	1288 (28%)
c) Plastic Glasses	7.5 (5%)	31.5 (15%)	264 (5%)	690 (15%)
d) Paper cups/Papers	15.0 (10.0%)	18.9 (9%)	516 (10%)	414 (9%)
e) Food wastage	4.5 (3%)	10.2 (4.80%)	152 (3%)	101.0 (3.08%)
f) Bins & Mise	15.0 (10.0%)	12.6 (6%)	524 (10%)	276 (6%)

Table 5-1: Waste Generation During Weekends—When Tourist Footfall Increases.[9] Source: Nature Climate Change

The hospitality and lodging industry has the fifth-highest rate of energy consumption in the world, expending energy in vast quantities to provide their guests with access to facilities such as heating, cooling, freshly cooked meals and more.

Water usage at these facilities is also a cause for concern, being used not only for drinking, cleaning, recreation and sanitary purposes but also for amenities, such as swimming pools and extensive landscaping.

Fragile Ecosystems

Many tourist activities, because of their alluring and exotic nature, take place in some of the most isolated,

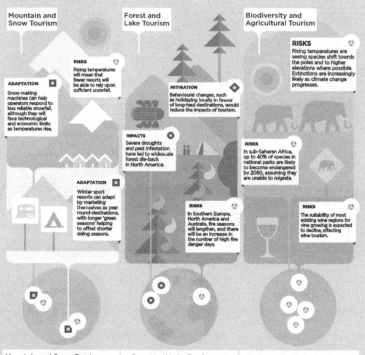

Tourism on the Move in a Changing Climate
Rising temperatures, higher sea levels and
degraded habitats will have serious impacts on
almost every sub-sector of tourism industry. But
options exist to help the industry adapt to
climate change.

IMPACTS Changes already affecting the tourism sector

RISKS Likely impacts on tourism in the future

ADAPTATION How the industry can respond

MITIGATION What tourism can do to reduce its emissions

Mountain and Snow Tourism

RISKS
Rising temperatures will mean that fewer resorts will be able to rely upon sufficient snowfall.

ADAPTATION
Snow-making machines can help operators respond to less reliable snowfall, although they will face technological and economic limits as temperatures rise.

ADAPTATION
Winter sport resorts can adapt by marketing themselves as year round-destinations, with longer 'green seasons' helping to offset shorter skiing seasons.

Forest and Lake Tourism

MITIGATION
Behavioural changes, such as holidaying locally in favour of long-haul destinations, would reduce the impacts of tourism.

IMPACTS
Severe droughts and pest infestation have led to widescale forest die-back in North America.

RISKS
In Southern Europe, North America and Australia, fire seasons will lengthen, and there will be an increase in the number of high fire danger days.

Biodiversity and Agricultural Tourism

RISKS
Rising temperatures are seeing species shift towrds the poles and to higher elevations where possible. Extinctions are increasingly likely as climate change progresses.

RISKS
In sub-Saharan Africa, up to 40% of species in national parks are likely to become endangered by 2080, assuming they are unable to migrate.

RISKS
The suitability of most existing wine regions for vine-growing is expected to decline, affecting wine tourism.

Mountain and Snow Tourism

Snow sports re t obvious risk from rising temperatures, with lower-elevtion resorts facing progressively less reliable snowfalls and shorter seasons. But other types of mountain tourism are also vulnerable, as infrastructure is put at risk from melting glaciers and thawing permafrost.

Forest and Lake Tourism

Outdoor activities will be affected by large-scale forest diebeck and more widespread wildfires, triggered by sustained drought and higher temperatures. Longer fire seasons will reduce access to national parks. Rising temperatures will change lake habitats, affecting fishing tourism.

Biodiversity and Agricultural Tourism

As temperatures rise, the geographical dispersal of flora and fauna will change, as species shift to conditions to which they are better adapted. Given that many nature reserves are geographically isolated, this my prove difficult or impossible for many iconic species.

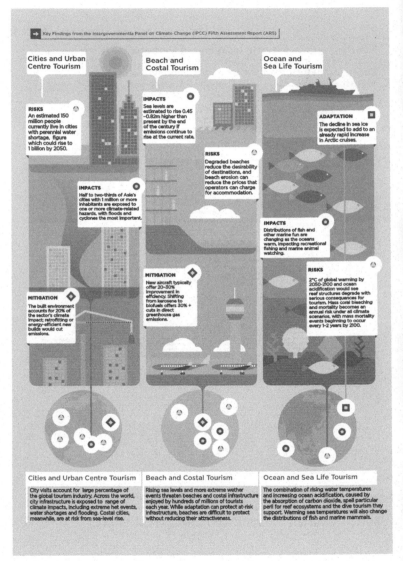

Figure 5-3: Tourism on the move in a changing climate[10]*. Source: University of Cambridge*

untouched and fragile ecosystems, such as mountain ranges, water bodies, villages and more. The very presence of tourists in non-controlled numbers can cause irreparable harm to rare flora and fauna native to the region.

For example, mountaineering tourists can harm the surrounding ecosystem by littering solid waste and trampling rare vegetation. Tourists indulging in activities like snorkelling and diving can also do more harm than they realize while exploring the ocean and other water bodies. While these activities are not necessarily harmful in and of themselves, there are a host of associated negative impacts that come with the presence of humans.

Malika Virdi, founder and CEO of Himalayan Ark, says:

'The Himalayan ecosystem is both geologically fragile and biodiverse. It cannot serve as a destination for mass tourism. Every year, approximately 7,00,000 people visit the region to seek out an adventure, or embark on a spiritual quest to find themselves. In some Himalayan regions, however, the local population is less than half that number. In such cases, the region cannot support the massive influx of people, along with their unsustainable practices. The way to go is slow and responsible travel.'

For example, coral reefs are highly sensitive by nature and can be severely harmed by the natural oils present on human skin. Accidental grazing, stepping on corals or boating activities that come into contact with coral reefs can be lethal to the corals themselves. In other cases, tourists may break off pieces of coral, or, in even more harrowing cases, the locals themselves will dynamite the reefs to sell pieces off to tourists.

These reefs are highly sensitive and take years to grow to maturity. Their destruction, on the other hand, can occur in a matter of seconds, dooming underwater ecosystems to a dismal future lacking in diversity and life.

Indigenous Fauna and Wildlife

While the industry has grave impacts on the environment and indigenous communities, there is another affected party without the voice to advocate for themselves, and who are often overlooked as a consequence—animals.

Animal-centric tourism accounts for 7 per cent of world tourism figures[11] and is growing at an annual rate of 15 per cent in Indian nature parks.[12]

While the objective of wildlife tourism is to observe the animals in the wild, the act of doing so is human-centric, with the animals being disregarded as living, breathing, sentient beings, instead perceived as attractions at a museum.

Private zoos are a great example of this disregard for the welfare of animals, as most recently demonstrated by the viral docu-series *Tiger King*. Wild animals are captured and removed from their natural habitats to be placed in cages for the amusement of others.

As pointed out by the documentary series, it is estimated that there are around 5,000 captive tigers in the US, more than the approximately 4,500 remaining in the wild globally.[13][14]

In other cases, wild animals are domesticated by cruel and torturous means and drugged into submission so that they may be presented to tourists at exotic petting zoos, where they are prodded and violated for the perfect instagrammable photo-op.

Even safaris are not without blame. Due to the rising popularity of safari tourism, there is an influx of human activity in protected areas that wild and endangered species call their home. While many safari organizations prioritize no-contact, sustainable and conscious tourism, others are laxer, with tourists leaving behind trash and other items in sensitive zones. In some cases, tour guides or tourists have been caught trying to feed the wild animals in order to get the perfect sighting, which disrupts the animal's natural hunting cycles.

Even responsible guides and tours have their pitfalls. The very presence of humans in these areas, no matter how limited or controlled, can be strange and uncomfortable to animals, who can become frightened or provoked at the sight and sound of jeep engines, camera flashes and human voices. The tracking of the animals also comes with dire consequences for the animals themselves. While tour guides and naturalists will track animal movement around protected areas to try and understand their hunting and nesting behaviours, the information gathered can be misused by poachers, who hunt and kill these endangered

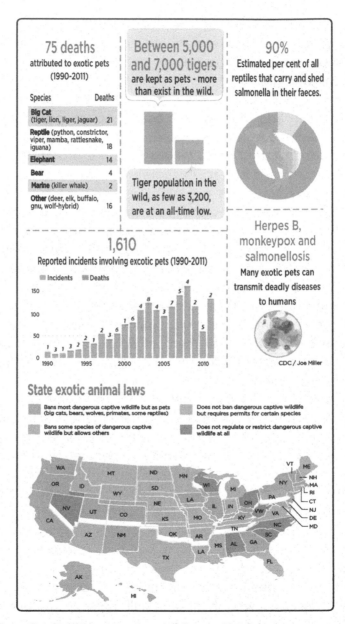

Figure 5-4: Statistics on exotic pets.[15] *Source: Live Science*

animals for trophies to be sold to the highest bidder on the black market.

Perhaps one of the biggest problems with commercial travel and international tourism is the travel itself. Every vacation comes with a carbon footprint, associated not just with the tourists' consumption habits, but with how they travel. Let's examine some of these in further detail.[16]

Air Travel

When was the last time you took a flight? Was it to meet a relative in a different city, state or even country?

Perhaps you were going on a long-deserved holiday to some exotic land?

Or maybe you were taking a work trip to a metropolitan city, only to return a few hours later

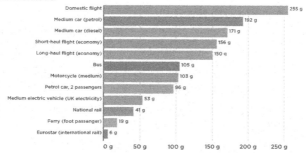

The carbon footprint of travel is measured in grams of carbon dioxide-equivalents per passenger kilometer. This includes the impact of increased warming from aviation emissions at altitude.

Domestic flight	255 g
Medium car (petrol)	192 g
Medium car (diesel)	171 g
Short-haul flight (economy)	156 g
Long-haul flight (economy)	150 g
Bus	105 g
Motorcycle (medium)	103 g
Petrol car, 2 passengers	96 g
Medium electric vehicle (UK electricity)	53 g
National rail	41 g
Ferry (foot passenger)	19 g
Eurostar (international rail)	6 g

0 g   50 g   100 g   150 g   200 g   250 g

Note: Data is based on official conversion factors used in UK reporting. These factors may vary slightly depending on the country, and assumed occupancy of public transport such as buses and trains.

Figure 5-5: Carbon footprint of travel per kilometre (2018).[17] Source: Our World in Data

after having signed and initialled some important documents?

Well, ladies and gentlemen, please fasten your seatbelts as we prepare for take-off, exploring the many ways air travel contributes to rapid climate change and unhealthy living conditions.

Aviation is responsible for approximately 2.4 per cent of global CO_2 emissions. When combined with other gases and water vapour trails produced by aircraft, the aviation industry contributes to about 5 per cent of total global warming.[18]

While international statistics indicate that few people qualify as frequent fliers, the wealthiest 10 per cent of the global population account for 76 per cent of the energy consumption associated with packaged holidays.

Let's take a look at fliers across the world:[19]

- US: 12 per cent of the population takes 66 per cent of the flights
- UK: 15 per cent of the population takes 70 per cent of the flights
- France: 2 per cent of the population takes 50 per cent of the flights
- Canada: 22 per cent of the population takes 73 per cent of the flights
- Netherlands: 8 per cent of the population takes 42 per cent of the flights
- India: 1 per cent of the population takes 45 per cent of the flights

The scary thing about this is that even considering the emissions from road transport, flying is the most damaging form of travel for our environment per kilometre.

In fact, as Leo Murray, director of climate charity Possible, puts it, 'Air travel is a uniquely damaging behaviour, resulting in more emissions per hour than any other activity, bar starting forest fires.'[20]

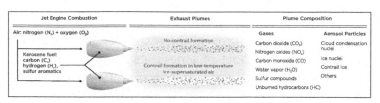

Figure 5-6: Pollution from jet engines.

While the aviation industry is working hard to combat its contribution to the damage, it is not moving nearly quickly enough, and with the numbers of flyers increasing year upon year, the gap is likely

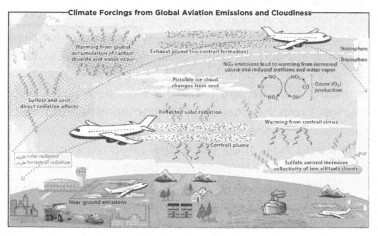

Figure 5-7: Climate forcings from global aviation emissions and cloudiness.[21] Source: Eureka Alert

to expand more rapidly than it can be bridged. The number of international travellers was expected to increase from 594 million in 1996 to 1.6 billion by 2020, adding greatly to the problem unless steps are taken to reduce emissions (WWF, 1992).[22]

Cruises

Luxury cruises are some of the most coveted vacation getaways. With a host of amenities on and off-board, cruises replicate a sense of community on the open sea, as well as a chance to interact with locals from the various ports on the cruise's route.

In recent years, the global cruise industry revenue grew to over 27 billion US dollars. Between 2017 and 2018, 138 cruise ships alighted on Indian ports, carrying a total of 1.76 lakh passengers.[23] However, the industry was severely impacted by travel restrictions associated with the coronavirus (COVID-19) pandemic in 2020, with numbers dropping sharply since.

Cruise companies and ships are notorious for their negative impacts on the natural world as well as local communities through means of air and water pollution, economic leakage and tax avoidance, as well as over-tourism. From the location of the company headquarters themselves to the routes the cruises take, as well as where and how they dock, the negative impacts of cruise travel carry through from beginning to end.

Cruise ship passengers tend to disembark all at once, which can cause overcrowding on small island shores.

Marine vehicles, such as cruise ships, harm not only the air quality of the region but also the surrounding water quality, due to their inefficient working and the release of fuel and ballast water into the ocean. From an environmental perspective, this means they release a range of air and water pollutants as well as harmful GHGs.[24]

In coastal regions and ports, nitrate, sulphate, particular matter and volatile organic compounds and deposits from ships and shipping activities can be significant, compromising air quality and affecting human health, leading to higher premature mortality rates. Estimates show that a single ship emits particulate matter equivalent to 100 million cars.[25] The fuel oil used by cruise ships contains about 2,000 times more sulphur oxide than ordinary diesel, according to the *Guardian*, causing pollution at ports.[26]

An average seagoing cruise produces 1.2 to 1.3 tonnes of CO_2 per passenger or around 169 kg of CO_2 per passenger per day. Total global CO_2 emissions from seafaring cruise ships have been estimated at 19.17 million tons in 2007, which is 1.5 per cent of global tourism emissions. Adding to this, the flights and associated travel emissions for the passengers and crew can effectively double this number.[27]

Emissions from international shipping are difficult to rectify, with few willing to take responsibility. Seafaring emissions are not regulated by the Kyoto Protocol because of the difficulties in allocating transnational emissions to any specific countries, and therefore, the onus is on the facilitating companies themselves. These companies will often avoid making costly changes to protect the environment if it means damaging their profit margin.[28]

The release of ballast water may also lead to the introduction of foreign species, which leads to increased competition among marine life and disturbs the natural ecosystem.

Another key issue that arises from commercial cruises is the environmental impacts associated with mooring activities, including anchoring, embarking and disembarking, re-supplying, and of course, the resulting recreational activities onshore.

These activities can result in a host of impacts, including a rapid uptick in local air, water and solid waste pollution, as well as damage to sensitive coral reefs.[29]

Adding to this, a significant portion of waste generated on the ships themselves is off-loaded at ports, and depending on whether or not they have been appropriately treated, could lead to a host of other problems for local systems. One person on a cruise produces about 2.6-3.5 kilograms of waste a day, compared to the average of 1-2 kg of waste produced on land.[30] This increase is associated with the use of SUP cutlery, crockery, disposable beverage bottles, travel wipes and more.

Automobiles

Much of the tourism-related air pollution comes from automobiles. They emit several toxic GHGs, including carbon monoxide, nitrogen oxide and volatile organic compounds. Exhaust from tourist cars, buses and other vehicles often affects the air quality and vegetation in national parks.

Road transportation accounts for about 11 per cent of India's carbon emissions and is a major source of pollution in several cities nationwide.[31] As many as fourteen of the twenty most polluted cities in the world are in India, according to the WHO.

What Are We Missing?

Travel has been romanticized by the social media monster, with scores of people looking to escape to distant lands in search of rejuvenation and a better understanding of self.

While your newsfeeds may be choked with colourful, exotic and eye-catching images of natural beauty and wonder, what many fail to notice is what lies just beyond the frame.

I have travelled to some of the most remote regions of India, places I have seen captured by thousands of the globe's most trending influencers and creators. However, what I found there was not what their pictures and stories would have me believe. Plastic is perhaps the most well-travelled pollutant of all, going everywhere human beings do—and sometimes reaching places we could never touch.

Plastic has seen the world and claimed it as its own—and we are the driving force behind its reign.

While travel is indeed the purest form of learning, teaching us not only about the world but also about ourselves, it must be undertaken consciously and in moderation.

Once, business meetings would entail multiple participants flying across states and borders under the premise of 'getting everyone in the same room'. That was until the travel restrictions associated with the COVID-19 pandemic shattered this myth, demonstrating the value of digital communication and highlighting the wasteful nature of age-old beliefs of how things are done.

To me, travel is the most direct route to higher wisdom, a chance to zoom out and expand your worldview. However, we must abandon our self-serving nature to allow ourselves to grow, expand our horizons and learn from our surroundings. As citizens of the world, we must realize that to truly learn from the natural world, we must first learn to treat it with respect.

Solutions: Conscious Travel

There is a multitude of ways to make travel more sustainable. Let's examine a few of these below.

Ecotourism

Ecotourism is about more than simply visiting natural attractions or natural places; it's about doing so responsibly and sustainably. It refers to a type of conscious tourism that provides a stable and sustainable means of income to local communities while respecting their traditions and practices, and protecting natural ecosystems.[32]

Shivya Nath, author and founder of Climate Conscious Travel, says:

'Conscious tourism is about the choices we make—from why we decide to go somewhere to how we get there, the food we eat to the way we interact with local communities, how we spend our money to where the benefit of that money goes, and how we choose to post about the experience on social media. On platforms like Instagram, the world has started to feel like one giant selfie backdrop. But what's the point of travelling if it doesn't change our perspective, or stir something deeper within us? Besides, the choices we make on the road also impact the places we visit and the people we meet. Conscious travel simply means recognizing that and choosing better.'

Ecotourism prioritizes reducing one's travel footprint while opening doors to the most authentic experience of the area. Here are some examples of ecotourism:

Accommodation

Staying in small-scale, family-run, local and traditional housing, preferably with renewable energy sources in place, can reduce the associated carbon footprint of tourist accommodation by as much as 48 per cent, as reported by the Global Footprint Network. They are also a great way to truly experience the local flavour, through the lens of the people that live there.[33]

For example, Ladakh is home to several affordable family-run homestays with backyard farming systems and rainwater harvesting equipment.

According to Ishita Khanna, co-founder and director, Spiti Ecosphere:

'Sensitive landscapes such as Spiti do not have the infrastructure to support vast numbers of seasonal tourists. To truly experience the area without becoming a burden upon its resources, I recommend that every tourist opt for Homestays over hotels. Homestays provide an insight into the area and the way the locals live. You are not a customer but a guest in someone's home and have the opportunity to develop real relationships and experience the culture of the area from a local lens. When you stay with locals, you directly benefit the community.'

According to a survey of Indian luxury travellers conducted in May 2020, a majority (68 per cent) of them stated they would pay extra for staying in hotels with eco-friendly and sustainable practices during the COVID-19 pandemic in 2020.[34]

Food

As per the Global Footprint Network, eating regional and locally sourced food improves each meal with a foo(d)print reduction of 5 per cent.[35]

Consider the quantity of food you are consuming to reduce the amount of food waste being generated during your visit. Eat light and balanced meals that prioritize local specialities and crops.

Travel Slow

Travelling 'slow' by using alternative and motor-free modes of transportation as much as possible can greatly affect your vacation's carbon footprint.

While walking for short distances is recommended, one can also travel longer distances on a bicycle to keep healthy while also keeping the health of our environment intact. After all, travelling by bike means not just a reduction in your travel footprint, but the eradication of it!

Carbon-Free Activities

Enjoying carbon-free activities, such as walking, cycling, kayaking, trekking and more, is a cost-effective, eco-conscious undertaking that both nature and your pocket will thank you for!

Ecotourism is slowly emerging as a model of sustainable travel in India. Ladakh, Uttarakhand, Sikkim and Meghalaya are just a few examples of states in which initiatives to promote sustainable tourism and carbon-free activities have been undertaken, facilitated by the help of local communities.

Aindrila Mitra Rajawat, editor-in-chief of *Travel & Leisure India*, says:

I believe there are two major factors to promoting sustainable tourism - Firstly, sustainability practices at a large scale can sometimes get expensive. So we need to be able to find a balance where organic and sustainable options can be available at a price conscious point. Secondly, of course, is awareness. Knowing the impact our visit can have on the awe-inspiring destinations we see can help us make more conscious decisions towards our travel.

In Bhutan, the government has taken steps toward eco-sensitive tourism practices by controlling the number of travellers to the area, and by promoting high-value tourism. Every visitor pays a minimum daily tariff of US$200 to US$250 (Rs 14,000 to 17,000), depending on the season.[36]

Reusable Travel Kits

Reusable travel kits are a great way to reduce waste generated on the road. Here's a helpful checklist of the must-haves in your travel kit:

- ☐ Reusable water bottles and mugs
- ☐ Eco-friendly food containers
- ☐ Reusable cutlery and straws
- ☐ Bamboo toothbrushes
- ☐ Travel towels
- ☐ Solid toiletries
- ☐ Eco-friendly sanitary products

Conscious Tourism Practices

If there was one piece of advice I would share with my fellow travellers, it is this: leave it as you found it.

Nature is powerful and equipped with all the tools it needs to protect itself. However, if we do not stop willingly causing it harm, it will turn on us until there will be nothing left to see.

Travel is a noble undertaking. It presents an unmatched opportunity to understand the world, and to learn how truly interconnected we all are. Travel represents the emotional over the material. It reminds us that we are only creatures, like any others on this planet. Like them, we must live simply, minimally, and with no attachments.

If there is one thing I have learnt from my life as a sustainable traveller, it is that your home isn't *your* home, your city isn't *your* city, your country isn't *your* country—the world is yours, and you are the world's.

There is no yours or mine in the larger world, only ours, and we must do all we can to protect what is ours, both for ourselves as well as for others.

Resources

Here is an inventory of supplementary media material, people who matter and shops in line with the chapter's theme.

Books

1. *The Shooting Star: A Girl, Her Backpack and the World*
    Author: Shivya Nath
    Published by: Penguin Random House India
    https://penguin.co.in/book/the-shooting-star/

2. *Sustainable Travel: The Essential Guide to Positive Impact Adventures*
    Author: Holly Tuppen
    Published by: White Lion Publishing
    https://www.quarto.com/books/9780711256019/sustainable-travel

3. *The Green Edit: Travel: Easy Tips for the Eco-Friendly Traveller*
    Author: Juliet Kinsman
    Published by: Ebury Digital
    https://www.penguin.co.uk/books/442279/the-green-edit-travel-by-juliet-kinsman/9781529107852

Documentaries

1. *Crowded Out: The Story of Overtourism*
    Directed by: Responsible Travel
    Available on YouTube
    https://www.youtube.com/watch?v=U-52L7hYQiE

2. *Eco India: An Eco Tourism Model that Focuses on Forest Conservation and Building the Local Economy*
    Directed by: DW and Scroll.in
    https://www.youtube.com/watch?v=_PL6tk110il

TedTalks

1. Sustainable Tourism - A Modern Eco-Friendly Perspective on Tourism
    By: Sumesh Mangalasseri
    https://www.youtube.com/watch?v=efgmEbjbGR8&t=97s

2. How 'Traveling Like a Local' Can Help Fight Overtourism
    By: Janek Rubes
    https://www.youtube.com/watch?v=36A5bOSP334&t=99s

3. 3 Ways to Make Flying More Eco-Friendly
    By: Ryah Whalen
    https://www.youtube.com/watch?v=JY-_GRi56KQ

People

1. Shivya Nath | Shivya
    https://www.instagram.com/shivya

2. Ketki Gadre | Ecokats
    https://www.instagram.com/ecokats

3. Ellie Cleary | Soultravelblog
    https://www.instagram.com/soultravelblog

4. Saravana Kumar | India in Motion
    https://www.youtube.com/c/indiainmotion

Shops/Stores

1. The Womans Company
    Eco-Friendly Feminine Hygiene Products
    https://www.thewomanscompany.com

2. Clan Earth
    India's most sustainable Backpacks
    Available across India
    https://clanearth.com

Sustainable Tourism Operators

1. Spiti Ecosphere
    https://www.spitiecosphere.com/

2. Himalayan Ark
    Eco-Friendly Homestays Group
    https://www.himalayanark.com/

3. Village Ways
    https://villageways.com/

4. The Blue Yonder
    http://theblueyonder.com/

5. Himalayan Ecotourism
    https://www.himalayanecotourism.com/

6

Tech and Digital

GHG Impact: 4 per cent[1]

This industry impacts the following Sustainable Development Goals (SDGs):
- SDG 6: Clean Water & Sanitization
- SDG 11: Sustainable Cities & Communities
- SDG 12: Responsible Consumption and Production
- SDG 13: Climate Action
- SDG 15 Life on Land
- SDG 14 Life Below Water

ACTIVITY

1. How many electronic devices do you use daily?
   - ☐ 1-2
   - ☐ 2-3
   - ☐ 3-4
   - ☐ Other (please specify)

2. How many of the following do you own?
   - ☐ Laptop
   - ☐ Mobile phone
   - ☐ Camera
   - ☐ Headphones
   - ☐ Smart Watch
   - ☐ Smart TV
   - ☐ Microwave
   - ☐ Audio & Video Equipment
   - ☐ Hard Drive
   - ☐ Massage Chair
   - ☐ Dishwasher
   - ☐ Heating Pad
   - ☐ Laundry Machine
   - ☐ Remote Control
   - ☐ Electronic Gym Equipment
   - ☐ Home Entertainment Devices
   - ☐ Gaming Console

3. How many hours a day do you spend on the internet for leisure?
   - ☐ 2-3
   - ☐ 3-4
   - ☐ 4-5
   - ☐ Other (please specify)

4. How many hours a day do you spend online for work?
   - ☐ 3-4
   - ☐ 4–5
   - ☐ 5-6
   - ☐ Other (please specify)

5. How many social media platforms do you use daily?
   - ☐ Instagram
   - ☐ Facebook
   - ☐ Twitter
   - ☐ Reddit
   - ☐ Other

Close your eyes for a moment and imagine yourself living in the early eighties. Imagine a simple life, with no concerns beyond tending to your home, your family and your work. Imagine then that someone told you that soon, computers—a radical invention—would change the way we live forever. From shopping to socializing, and from dating to the global economy, imagine someone telling you that in the not-so-distant future, all people across the globe would exist in a state of perpetual connectivity, bound together by invisible threads that form a globe-sized network of information.

Imagine that someone told you that you would soon own a handheld device, no larger than the palm of your hand, more powerful than any supercomputing device, which could contain more information than the whole Library of Alexandria, and that you would use it, primarily, to look at pictures of cats.

Without a doubt, this would seem like an absurd notion concocted by the town fool—but then, of course, it happened.

Within the span of just a few years, science fiction transformed into our reality, and we were propelled into a future only dreamt of by scientists and writers of ages past.

Where once correspondences would be scribbled on paper and carried across the land and the sea to their intended recipient by hand, ship or pigeon, today, email is considered the standard. Documents, files and even music travel from one end of the world to the other in milliseconds, quicker than even the blinking of an eye.

From entertainment to work, the digital age has changed the way we live forever. Our quest for convenience has thrust us into a future where the impossible is made possible with the click of just a few buttons, and where final frontiers seem less and less final every day.

Today, the information technology (IT) market size is expected to reach $13,818.98 billion in 2026 at a compound annual growth rate (CAGR) of 10.3 per cent, with the global digital transformation market size expected to grow at a CAGR of 21.1 per cent, to reach USD 1,548.9 billion by 2027 from USD 594.5 billion in 2022.[2]

Analyses of trends in employment reveal that the tech workforce in 2020 was 5.5 million strong and has been growing at an average annual rate of 2.2 per cent since 2001. The pace of growth is well above the 0.4 per cent average rise per year for total employment in the economy during this period.[3]

The astonishing thing about the world we live in today is how quickly it evolves. We are currently standing on the brink of the age of Web 3, which promises to unite people, economies and countries under the banner of a decentralized World Wide Web—and the best part? We're not done growing yet.

If our ancestors could see us now, they would label us witches and heretics, claiming we had defied the natural order of things and flown in the face of whatever gods they believed existed.

But what if they were not wrong?

I am a digital creator. My life's work is a repository of films, blogs, photographs and more that would not have been possible without the use of technology, the internet and my many electronic devices. The internet and technology have irrevocably changed my life—and for the better. After all, without social media, the internet, my computer, cameras and a plethora of recording devices, I wouldn't be here, writing this book today.

I am known as a digital nomad, meaning that I have no home besides the one I create for myself. I have no community, besides the invisible web that connects me to the thousands of people I reach every single day. But what happens if, for some reason, I can no longer connect with these people?

When things evolve, they leave behind a trail of outdated ideas and technologies that just couldn't cut it in the new world. As such, I am acutely aware of the inertia of time and how it makes way for the new while relegating the old to the recesses of distant memory and nostalgia. As a digital creator, I too am one of these shiny, new objects, waiting to be replaced by a person, a device, a technological innovation or some combination thereof that does what I do better than me.

But the problem with the digital and ICT industry is larger than me—while social media would have us follow the words and actions of the singular over the collective, we must remind ourselves that it is not about the individual, but about all of us.

Let me tell you about the hidden evils of the tech industry, and how you and I are participating in a race for a future we may never get a chance to fully experience.

Worldwide Wish-Fulfilment—The Problem with ICT

In Germany in the year 1440, a goldsmith named Johannes Gutenberg invented the printing press, changing the way humans communicated with one another forever.

Since then, human society has undergone several technological revolutions. From the age of analogue telecommunications to the birth of the World Wide Web, technology has revolutionized the way we think, create, communicate and connect.

Today, path-breaking inventions are the everyday currency of tech hubs like Silicon Valley, with a new one hitting the stands every other day. Tech giants release new models of laptops, phones, televisions, gaming consoles, smartwatches and more every year. Every release is accompanied by thousands of tech enthusiasts lining outside storefronts to get their hands on the latest edition before their friends. The US Environmental Protection Agency estimates that over 438 million electronic devices were sold in 2009, which is twice the amount sold in 1997.[4]

But have you ever paused to think about what it takes to produce new gadgets at such vast volumes, year upon year?

Manufacturing new gadgets and electronic equipment requires a host of expensive, rare and environmentally taxing raw materials.

From plastic to heavy metals, lead, batteries and more, the manufacturing of hardware involves widescale mining practices for raw materials such as copper, lithium, cobalt, manganese, nickel and graphite, among others, leading to the clearing of areas of sensitive forested land. This poses a serious threat to the flora and fauna of the region.

These raw materials are expensive and environmentally taxing to extract and are sourced from non-renewable deposits deep within the earth's crust. As such, mining practices, such as explosions and more, result in vast amounts of chemical run-off, loosening of soil as well as damage to the health of miners and surrounding communities.

The raw materials then must be transported to manufacturing hubs, where they are processed and made into parts appropriate for use in gadgets. These processes are numerous, and typically generate a lot of waste, further deteriorating the environment. Following this, the product must be assembled through the use of large, industrial-scale machinery, before being packaged and prepared for delivery.

When the finished devices hit the market, they are transported to consumers across the globe, further adding to their carbon footprint. This cycle then repeats itself year upon year.

But what is the appeal of a new model of the same device released within such a short period? Electronic devices are equipped with a host of new and exciting features to attract returning and potential customers, but one of the biggest appeals of upgrading a device is storage.

Today, a gadget that can be worn around your wrist is equipped with the capacity to store multitudes of gigabytes of information that our ancestors would

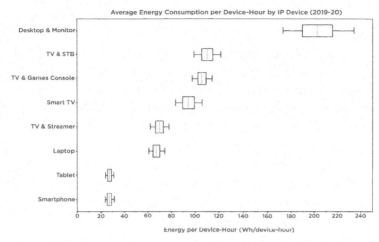

Figure 6-1: Average energy consumption per device-hour by IP device.[5]
Source: BBC

never have dreamed of containing in a single volume of text. But how is that possible?

While most devices today are equipped with more local storage than the floppy discs and external storage devices of ages gone by, most of our information is stored in the cloud—an invisible, wireless storage facility that allows users to access their files from any device, anywhere in the world. This also allows users to store less information locally, which means that their data won't slow down the operating functions of their devices.

However, while the cloud is indeed an invisible convenience, it also has its invisible costs to our environment. Cloud computer sites can consume up to 622.6 billion kilowatts per hour of power, consuming power resources in vast volumes to satisfy our computing needs. Estimated cloud consumption accounts for 1–2 per cent of the world's electricity resources.[6]

While cloud storage does indeed take up a lot of power, the ecological impact of the cloud is a highly debated notion, with some claiming that it aids in saving up physical resources. However, there is a small section of experts who have criticized cloud storage for consuming high megawatts of power.

Although the energy consumption is high, running business applications in the cloud is widely considered to be more energy- and carbon-efficient than operating via on-premises server rooms.

Nikhil Arora, thought leader and CEO, says:

> 'To thrive in the face of escalating concerns about climate change and environmental impact, leaders must prioritize tracking and addressing environmental risks and opportunities while integrating the triple bottom line–people, planet, and profits–into decision-making. Effective leadership in sustainability is indispensable for fostering a culture of success.'

Many major brands have taken to greening their clouds by committing to 100 per cent renewable energy. Some of these include Meta[7], Google[8], Apple[9] and Microsoft.[10]

However, information stored on the cloud must have a physical counterpart, known as a server. Servers are complex computing machines that are typically housed together in commercial data centres. The Asia

Pacific region is home to the most cloud data centres at ninety-five, with the United States and Canada region not far behind with seventy-nine centres. Together, these regions account for 72 per cent of the world's cloud data centres, with Europe housing 24 per cent and Latin America just 4 per cent.[11]

But what exactly are these data centres?

Data centres are responsible for the housing of computer systems, network and data storage. Data centres store information on servers located in vast data housing facilities around the world and enable the delivery of shared applications and data to users worldwide.

The production, shipping, and maintenance of these servers is an energy-intensive process for a multitude of reasons. For a data centre to remain optimally functional, it needs to be built in a cold region or a temperature-controlled environment. According to studies, around 40 per cent of the total energy that data centres consume goes towards cooling IT equipment.[12]

With the launch of 5G, the new wave of IoT devices, and a thriving cryptocurrency movement to add to the problem, data centres are becoming more and more necessary to store and process our data. As more devices become connected, more data will need to be processed than ever before.

These data centres expend high amounts of energy, most of which is sourced from the use of fossil fuels, contributing to GHG emissions. It has been said that data centres contribute around 0.3 per cent to overall carbon emissions, and use an estimated 200 terawatt hours (TWh) each year, more than the annual energy consumption of some countries.[13]

To illustrate this, let's take the example of the popular remix track *Despacito* by pop icon Justin Bieber. When the music video was first released, it became an instant sensation and earned the title of the first video to hit 5 billion views on YouTube. Scientist Rabih Bashroush calculated that the associated downloads consumed as much energy as Chad, Guinea-Bissau, Somalia, Sierra Leone and the Central African Republic put together in a single year. Today, that same video is on the precipice of crossing 8 billion views.[14]

The entire information technology (IT) sector—from powering internet servers to charging smartphones—is estimated to have the same carbon footprint as the aviation industry's fuel emissions.[15]

Another important function that servers provide is housing information that can be accessed via the internet. On 1 January 1983, the internet was born, completely changing the face of human existence as we know it. Before this date, computer networks had no way of connecting with each other.

To connect to the World Wide Web, a device must be equipped with network access. Network access allows users to dial into the internet through various means, with Ethernet and Wi-Fi being two of the most popular.[16]

Network access through local area networks (LANs) refers to network access at residential, commercial, educational and public facilities, and is area-bound.

LAN is facilitated by energy consumption and carbon emissions that are produced via two major

THE CARBON FOOTPRINT OF THE INTERNET

CO_2 EMISSIONS FROM EMAILS, SEARCHES, AND CLOUD STORAGE

SIZING UP THE INTERNET'S CARBON EMISSIONS

CARBON FOOTPRINT

4 BILLION+
Over 4 billion people are active internet users.

3.7%
The carbon footprint of our gadgets, the internet and systems supporting them accounts for 3.7% of global greenhouse emissions, similar to the airline industry. These emissions are predicted to double by 2025.

NO. 3
Global IT sector electricity demand ranks behind only two countries in the world- China and the US.

THE INTERNET USES A HUGE AMOUNT OF ENERGY. THIS IS DUE TO TWO KEY FACTORS:

MANUFACTURING AND SHIPPING

Technology companies must manufacture and ship the internet's hardware globally.

POWERING AND COOLING

Computers and smartphones must be powered and cooled, drawing electricity from local grids.

SEARCH ENGINE

SERVER

WEB BROWSER

EMAIL

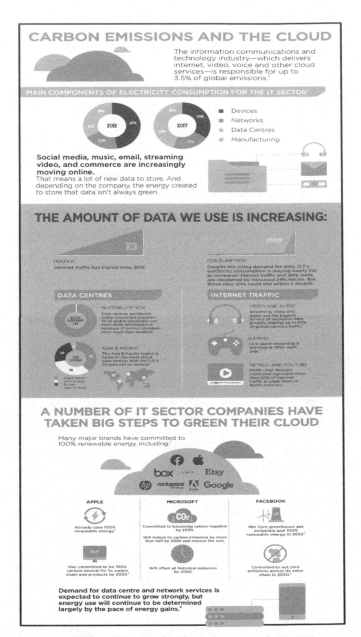

Figure 6-2 and 6-3: The carbon footprint of the Internet.

stages of providing access. The first is the router manufacturing stage, which is responsible for up to 62-80 per cent of total energy consumption. The second is the annual energy consumption of operation, which accounts for 20-38 per cent.

Regardless of whether you use a Wi-Fi network or cellular data, more than half of the world's population today is online. Internet traffic has almost tripled since 2015, and with it, the need for and consumption of data stores has increased too.[17]

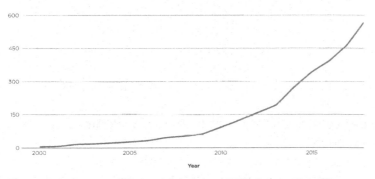

Figure 6-4: Internet traffic has tripled since 2015. Source: Reset[18]

A report by Greenpeace rated various platforms based on their energy transparency, renewable energy commitment, efficiency, greenhouse gas mitigation, RE procurement and advocacy, and found that streaming video and audio are the biggest drivers of explosive data growth, and make up 63 per cent of the global internet traffic.[19]

According to the report,[20] live game streaming is growing by over 19 per cent each year, and Netflix and YouTube combined are responsible for more than 50

Internet Company Scorecard

Video Streaming

	Final Grade	Clean Energy Index	Natural Gas	Coal	Nuclear	Energy Transparency	Renewable Energy Commitment & Siting Policy	Energy Efficiency & Mitigation	Renewable Procurement	Advocacy
Afreeca.com	F	2%	19%	39%	31%	F	F	F	F	F
Amazon Prime	C	17%	24%	30%	26%	F	D	C	C	B
HBO	D	22%	20%	25%	25%	D	F	F	F	F
Hulu	F	20%	30%	29%	20%	F	F	F	F	F
Netflix	D	17%	24%	30%	26%	F	F	C	D	F
Pooq.co.kr	F	2%	19%	39%	31%	F	F	F	F	F
Vevo	F	27%	15%	32%	26%	F	F	F	F	F
Vimeo	D	47%	13%	20%	19%	D	F	F	C	F
Youtube	A	56%	15%	14%	10%	B	A	A	A	A

Figure 6-5: Internet company score card.[21] Source: Greenpeace

per cent of the internet traffic at peak times in North America.

Let's look at some of the ways our digital consumption patterns affect our environment, and measure how some of our favourite service providers fared when graded on their ethics and sustainability:

Video Streaming

Amazon Prime Video: C grade
HBO: D grade
Netflix: D grade
YouTube: A grade[22]

From YouTube to Netflix, video streaming has transformed the entertainment landscape irrevocably.

Figure 6-6: Carbon dioxide produced by Internet activities daily.[23] Source: Geneva Environment Network

OTT platforms have replaced movie theatres due to their accessibility, allowing users to stream their favourite movies and TV shows from any device, anywhere in the world.

And it's not just entertainment, either. During the COVID-19 pandemic of 2020, video calling and remote work became the norm, with thousands of people streaming live feeds of classes, meetings and chats with their friends around the globe daily. While some harbour a belief that streaming lower resolution videos is a sustainable move for the environment, the energy consumption and carbon impacts of an hour of video streaming, whether in SD, full HD or 4K, will only be marginally greater for the higher quality content.

In 2020, there were over three billion internet users across ages who watched streamed or downloaded video via any device at least once per month. This figure is projected to increase annually to reach nearly 3.5 billion by the year 2023.[24] The video streaming market value amounted to 59.14 billion in 2021.[25]

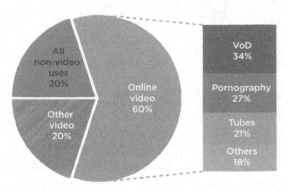

Figure 6-7: Distribution of online data flows between uses of digital technologies and of online video in 2018 in the world.[26] Source: Research Gate

While the convenience of streaming videos is an unparalleled luxury, we have more to worry about than slow buffering rates. With every video streamed, a certain amount of CO_2 emissions are released into our atmosphere. Compounded over time, and via multitudes of people streaming simultaneously, this seemingly innocent pastime can spell doom for our planet. Online video viewing, which represents 60 per cent of the world's data traffic, generated more than 300 $MtCO_2e$(Metric tons of carbon dioxide equivalent) during 2018. This is the equivalent of the annual emissions of Spain.[27]

To put this into perspective, watching a half-hour show leads to emissions of 1.6 kilograms of carbon dioxide equivalent. That's equivalent to driving 6.28 kilometres (3.9 miles).[28]

Take a look at this chart:

- The report[29] published by Germany's Federal Environment Agency calculated the amount of carbon dioxide produced per hour by data centres for streaming using:
  o Fibre optic cables result in 2 grams of CO_2 emissions
  o Copper cables produced 4 grams of CO_2 emissions
  o 3G mobile technology results in a hefty 90 grams of CO_2
  o 5G would result in carbon dioxide emissions of 5 grams of CO_2

Another taboo topic that contributes to global streaming numbers is pornography. Streaming online pornography produces the same amount of carbon dioxide as all of Belgium. Reports show that overall, online videos emit 300 million tonnes of carbon each year—a third of which comes from streaming videos with pornographic content alone.[30]

Figure 6-8: Streamed video and Internet surfing compared to electricity consumption of other activities.[31] Source: Ericsson

Author and doctor, Tanaya Narendra says:

'I believe every industry has a responsibility to make itself greener—and the porn industry, which can be particularly damaging to the environment, is no exception. While one might assume that digital pornography has made things better through the reduced consumption of paper, CDs, etc. some researchers are concerned that the ease of digital pornography may have been even more detrimental— pornography is so widely and easily accessible on the internet, that we have an unprecedented number of viewers tuning in to porn nowadays, generating loads and loads of carbon. Adding to this, the pornography industry is also closely linked to other illegal industries that fuel crimes towards humanity, such as human trafficking, illegal arms trading, and even global terrorism.'

Music Streaming

iTunes: A Grade
Spotify: D Grade
Soundcloud: F Grade[32]

Music is perhaps the only universal language, appreciated and enjoyed by billions across the globe every day. Where once CDs and cassette tapes ruled the market, contributing to rampant plastic waste generation, today, music streaming is the preferred form of consumption.

And it's not just music that these platforms are known for! From audiobooks to podcasts, the audio entertainment industry has found its roots in the digital age and is expanding at a rate never before witnessed across history.

The music industry is also taking direction from the age of social media and consumption, with artists and labels vying for the next viral TikTok track to break the internet with. According to estimates in the 2014 book, *The Evolution and Equilibrium of Copyright in the Digital Age*, music releases per year numbered 40,000. However, current estimates are more than double that number, at 1,00,000.[33]

However, a recent comparison of the carbon emissions of the two means of listening to music showed that emissions due to streaming are much higher.[34]

While GHG emissions from the music industry peaked at 157 million in 2000 under the physical era, the generation of GHGs via the storing and streaming of digital files is estimated to be between 200 million kilograms and over 350 million kilograms in the USA alone.[35]

Search Engine

Google: A Grade
Yahoo: B Grade
Bing: B Grade[36]

In the modern world, the word 'Google' is used as a verb that means 'look up on the internet'. Google and other search engines have provided us access to the whole gamut of humanity's knowledge—but this is not without its cost.

Internet searching is a tricky matter, speaking from an environmental perspective.

According to an estimation by FastCompany[37], Google.com weighs 2 MB and processes about 47,000 requests every second; the page emits 500 kilograms of CO_2 emissions per second. That's 300 tonnes of CO_2 every minute.

The carbon impact of the 80,000 requests made in 1 second (assuming the requests made were non-complex in nature and launched from a mid-range smartphone) worldwide is 8,660 $gEqCO_2$ (gram equivalent of CO_2), or the equivalent of 77 km travelled in a light vehicle.

The carbon impact of a day of Google queries is the carbon equivalent of 6.7 million km in a light vehicle.[38]

Figure 6-9: Carbon impact of basic search vs URL.[39] *Source: Greenspector*

While a decade ago, each internet search had a footprint of 0.2g CO_2e,[40] according to figures released by Google, today, Google uses a mixture of renewable energy and carbon offsetting to reduce the carbon footprint of its operations.[41]

On the other hand, Microsoft, which owns the Bing search engine, has promised to become carbon negative by 2030.[42]

Messaging

iMessage: A Grade
WhatsApp: A Grade
Skype: B Grade
WeChat: F Grade[43]

One of the biggest perks of the Internet in today's world is its ability to connect people.

SMS text messages are considered the most environmentally-friendly medium for messaging, with each text generating just 0.014g of CO_2e.[44]

According to the BBC, Messenger is only slightly less carbon intensive than sending an email.[45]

Social Media Platforms

Facebook: A grade
Instagram: A grade
LinkedIn: B grade
Pinterest: F grade
Twitter: F grade
Reddit: F grade[46]

Social media platforms have become digital repositories of our lives, our thoughts, actions and memories—so

Carbon impact of Instagram features
The lower this value, the better the functionality!

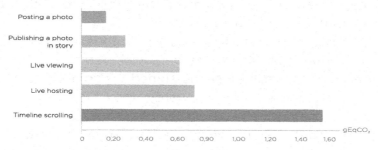

Figure 6-10: Carbon impact of Instagram features.[47] Source: Greenspector

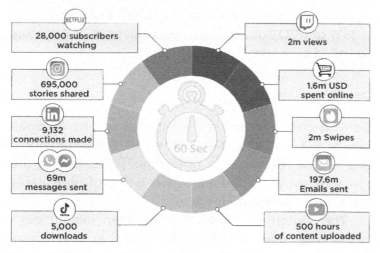

Figure 6-11: Estimated data created on the Internet in one minute.[48] Source: Statista

much so that 4.48 billion people currently use social media worldwide, a number that has more than doubled from 2.07 billion in 2015.

According to studies, a person spends an average of 2 hours and 24 minutes on social media in a day. This means that if someone signed up on a platform at sixteen and lived to seventy, they would spend approximately 5.7 years of their life online.[49]

However, every single second spent on social media contributes not only to emissions associated with streaming and data use, but also affects the amount of data created.

This creation of data implies a need to record and store it, as well as the functionality for consumers to access it from wherever they are across the world.

The average impact of a mobile Instagram user is 18.6 $gEqCO_2$/day, i.e., the equivalent of 166 metres travelled by a light vehicle.

Reading

Do you consider yourself a bibliophile? As all book lovers will tell you, the battle between traditional and digital reading is a polarizing subject, with the romanticism of traditional books being of undeniable appeal to readers of all ages.

The switch to e-readers has been lauded as a sustainable move, owing to the fact that fewer trees need to be cut down to produce volumes by the page.

However, as this chart confirms, this is not really the case.

Estimated emissions per unit

Newspaper	0.62 kg
Magazine	0.95 kg
Book	7.46 kg
iPhone	55 kg
Kindle	168 kg

Table 6-1: Estimated emission per unit. Source: Industrial Design Consultancy, Babcock School of Business, U.S., Environmental Protection Agency, Green Press Initiative, Marmol Razinder Prefab, Discovery Magazine, Apple, Cleantech Group analysis[50]

The requirement of raw materials and the process of manufacturing an e-reader is an energy and water-intensive process, with one e-reader being the equivalent of forty or fifty physical books.[51] Moreover, the emissions created by a single e-reader, including its production and the energy needed to charge it, equal about 100 physical books.

Adding to this, the adoption of e-readers themselves has been a slow process, with printed volumes still outnumbering e-reading devices around the world.

On the other hand, audiobooks have a lower carbon footprint overall, owing to the lack of any production of physical components. While audio files are indeed heavier than text files, being more data-heavy and requiring more storage space, they require no shipping or transportation, compensating for their storage-related emissions.

Cryptocurrencies

Cryptocurrencies have become all the rage in the years following 2020. A democratic, secured economic

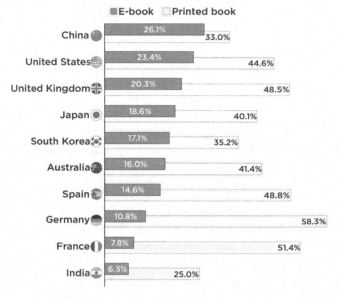

E-Books Still No Match for Printed Books

Estimated share of the population that purchased an e-book/a printed book in 2021

Figure 6-12: Estimated share of the population that purchases an e-book/a printed book in 2020.[52] Source: Statista

revolution in the making, cryptocurrencies have attracted the attention of Wall Street professionals and internet trolls alike, promising a brand-new horizon for both information technologies and economies at once.

In 2019, the global cryptocurrency market was approximately $793 million and is now expected to reach nearly $5.2 billion by 2026, according to a report.[53] Between July 2020 and June 2021, the global adoption of cryptocurrency surged by more than 880 per cent—an unmatched figure for a single year.[54]

Ashish Singhal, co-founder and CEO of Coinswitch, says:

> 'In the case of Bitcoin, energy consumption is not a design flaw but a feature. The energy consumption is the cost of securing the network. The more the number of miners, the more secure the network is. This, naturally, means higher energy consumption. There is, of course, room for improvement. Bitcoin mining is price-sensitive, interruptible, adjustable, and location-agnostic. These factors make it an ideal market for renewable energy. There is a clear sign that projects are increasingly focussing on becoming greener because the markets are rewarding greener projects. This is similar to how the equity markets reward companies with better ESG norms.'

The rise of cryptocurrencies has also birthed a whole new type of industry, that of commercial cryptocurrency exchanges. Coinbase[55], Binance[56], Bitfinex[57] and more, which did not even exist a few years ago, have transformed into industry leaders already.

However, cryptocurrencies consume massive amounts of energy, making them unsustainable for our long-term ecological health.

To understand how this happens, we must first understand what it takes to create a cryptocurrency. Unlike regular money, which is monitored and regulated by banks, transactions in cryptocurrency are tracked using blockchain, a publicly accessible, tamper-proof ledger facilitated by a network of computers around the world.

These transactions are verified via a process known as mining. Mining refers to a process in which computational puzzles are solved to verify transactions between users, which are then added to the blockchain. Besides being an inherently complex undertaking, mining is an energy-intensive process.

Take Bitcoin, for example. According to several calculations by the Bitcoin Energy Consumption Index, a single Bitcoin transaction consumed around 819kWh.[58] Comparatively, the same amount of energy could be used to operate a 150-watt refrigerator for about eight months. Further, a study by the Technical University of Munich determined that the entire Bitcoin system produced around 22 megatonnes of carbon dioxide per year, twice that of Varanasi's annual carbon footprint.[59] [60]

However, when compared to traditional banking systems, cryptocurrencies still present a more sustainable alternative to economics.

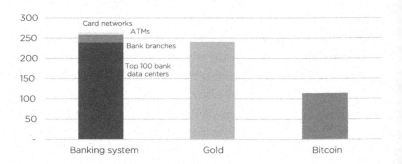

Comparing the energy consumption of Bitcoin, gold, and the banking system

Figure 6-13: Estimated annual energy consumption.[61] Source: Galaxy Digital

Email

In terms of its sustainability, emails are less taxing to our environment than the simple act of sending a post to a friend on Instagram or Facebook. The true problem with emails, however, occurs when they are sent gratuitously.

One email produces an estimated 0.000001 tonnes of CO_2 equivalent.[62] While that number is not an impressive one, when one considers the number of unnecessary emails sent and received in just a week, that number compounds very quickly. Every day, 319.6 billion emails are sent and received.[63]

A hundred short emails with an attached document of 1 megabyte generate as much electricity as watching a 10-minute video film.[64]

How Bad are Bananas?: The Carbon Footprint of Everything, published in 2010 by Mike Berners-Lee,[65] a researcher on carbon footprinting shows:

1) Spam email: 0.3g CO_2
2) Regular email: 4g CO_2
3) Email with attachment: 50g CO_2

Currency	PJ/y
Bitcoin	3.97
Paper currency and minting	39.6
Gold mining	475
Gold recycling	25
Banking system	2340

Table 6-2: Environmental costs of bitcoin mining compared with other systems.[66] *Source: ResearchGate*

Country-wise emails sent—Data from Statista:

1) United States: 9.6 billion emails (total), 8.6 billion emails (spam)
2) Germany: 9 billion emails (total), 7.9 billion emails (spam)
3) China: 8.7 billion emails (total), 8.5 billion emails (spam)
4) United Kingdom: 8.6 billion emails (total), 7.7 billion emails (spam)
5) Russian Federation: 8.4 billion emails (total), 8.1 billion emails (spam)

- Yearly amount of emails sent = 320 billion x 365 days = 117.8 trillion
- Yearly amount of spam emails sent = 117.8 trillion * 0.92 = 108.4 trillion
- Yearly amount of regular emails sent = 117.8 trillion * 0.08 * 0.76 = 7.2 trillion
- Yearly amount of emails with attachment sent = 117.8 trillion * 0.08 * 0.24 = 2.3 trillion

The yearly CO_2 emissions per email type:

	number of emails in trillion	CO_2 in million tonnes
Spam	108.4	32.52
Regular	7.2	28.80
Attachment	2.3	115.00

Compounded, these numbers mean an overall CO_2 emission of around 176.3 million tonnes per year from emails alone. This is the equivalent of the energy consumption of 21.2 million homes for one year, or 21 trillion smartphones charged, or the equivalent of a car driving 442.3 billion miles.

While these numbers are indeed shocking, compared to global CO_2 emission associated with fossil fuels (which was nearly 35 billion tonnes in 2020), our emails account for only 0.5 per cent of GHG generation.

E-waste

Digital consumption is an intangible force, one that is invisible to the naked eye while having some serious effects on our health and on our planet's ecological integrity. However, there is a more tangible phenomenon that results from our excessive reliance on technology and digital apparatus—one we can see with our eyes, hold in our hands and feel across every corner of the world: e-waste.

Electronic waste is a seldom discussed issue that has presented itself as an emerging cause of concern for both developing and developed countries worldwide. Due to the complex nature of the manufacturing of electronic devices, improperly disposed devices contain a multitude of parts and components made with harmful chemicals and toxic materials that naturally leach from the metals inside when buried, which can have negative effects on both our health as well as that of the environment. These parts also contain valuable, unrenewable resources that, if properly recycled, can be reused.

With global access to electronic devices improving day by day, and with the accelerated rate of new releases and consumers' use-and-throw mentality, e-waste is becoming a more pressing issue every day. Global electronic waste generation reached a record high of 53.6 million metric tonnes in 2019—an increase of 21 per cent in just five years, i.e., approximately 7.3 kilograms of e-waste per capita.[67]

While you are probably imagining heaps of defunct laptops and mobile phones piled up in landfills around the globe, in reality, e-waste is a far-reaching issue that comprises just about any discarded electrical or electronic equipment. From unsold items at stores to items donated to charity, e-waste has many sources, and just as many associated risks.

Virtually all electronic waste contains some form of recyclable material, but the process of recycling these materials is more difficult than it may appear. According to a 2019 United Nations report, consumers discard 44 million tonnes worth of electronics each year, only 20 per cent of which was recycled sustainably.[68]

E-waste management is a complex and costly endeavour that involves multiple stages, with numerous sub-stages for each. These include collection, segregation, dismantling, refurbishing and recycling. China is the largest producer of e-waste in the world, followed by the US and then India.[69]

In India, the collection, recycling and disposal of e-waste present an especially pressing challenge due to the lack of appropriate infrastructural capabilities.

According to a Central Pollution Control Board report, in the financial year 2019-2020, India generated 1,014,961.2 tonnes of e-waste under twenty-one categories of electrical and electronic equipment (EEE). Currently, more than 95 per cent of this waste is handled by the informal sector, in which people often work without the knowledge, equipment or safety precautions needed, only adding to the problem.[70]

Ghana, China, India and Nigeria serve as the dumping grounds of most of the e-waste produced by developed countries. This in turn causes widespread suffering and environmental degradation.[71]

The workers who sift through the guts of everyday technology lack even the most basic safety equipment, such as gloves and goggles, but are exposed to a number of harmful chemicals, including, but not limited to, arsenic, lead and mercury. Several electronic gadgets also contain significant amounts of lead and other toxic and carcinogenic compounds. Long-term exposure to these substances can damage the nervous system, kidneys, bones and the reproductive and endocrine systems.

In India, metropolitan cities such as Bengaluru, Delhi and Mumbai face detrimental impacts due to growing e-waste generation. With over 1,200 foreign and domestic technology firms, Bengaluru is a city that figures prominently on the list of cities endangered by e-waste-associated health hazards. According to reports, as many as 1,000 tonnes of plastic, 300 tonnes of lead, 0.23 tonnes of mercury, 43 tonnes of nickel and 350 tonnes of copper are generated annually in Bengaluru.[72]

Improper disposal of e-waste also contaminates the environment when incinerated or buried in landfills along with domestic waste. Natural processes, such as leaching, can lead to acidification of surface waterways, and increased levels of lead, mercury, zinc and other harmful elements in landfills affect surrounding soil, while incineration of certain chemicals and materials leads to the release of toxic fumes and GHG emissions.

What Are We Missing?

We are products of the age of hyper-connectedness, a period critical for humanity's survival. In this age, being off the grid is more a test of our emotional, mental and physical strength than it is a lifestyle choice.

I am no exception. My career—even my life—is entirely dependent upon the internet and my electronic devices. Without them, I am just another nomad, asking passers-by to heed my warnings about environmental sustainability.

I owe much of my development to the internet. As a child disillusioned by the education system, the internet became my teacher. My greatest mentors, the Dalai Lama, Bhagat Singh and so many more, spoke to me directly through my computer screen, blurring the lines between the past and the present, between the living and the dead, between here and there. All the people I know, love and cherish—from my friends and my support system to my professional network—first came to me through the internet.

The internet has been a constant companion on my journey, and we have grown and evolved together. I began my career writing blogs, documenting my travels around the nation. However, as internet preferences changed from the written word to the visual, I too followed suit, foraying first into photography, and then to video. Even my few offline projects rely on the internet and digital media for their promotions.

The internet has made me a global public figure from the comfort of my sleeping bag. I have never circumnavigated the globe, or seen distant lands with my own eyes, but the internet has allowed me to connect with people across oceans and seas and has opened my eyes to the larger world out there—as I imagine it has done for many of you.

However, the digital world is not as rosy as it may seem on the surface. As with most others, the digital and electronic industry is a capitalistic structure that prioritizes profits over all else. While the World Wide Web is considered the most democratic, tangible example of free speech and free will, we have far less freedom than we are led to believe.

Social media companies and developers influence our tastes every day, causing a ripple effect that reaches billions across the globe. However, when this influence is inauthentic, manufactured by the world's capitalistic giants, it can be disastrous for those who fall prey to their narratives.

Creators are being forced to diversify and change the way they operate to get paid and gain the largest number of views and followers to stay relevant. This means compromising their ideals in exchange for petty cash and 15 seconds of fame. Companies involved in greenwashing and other unethical practices manipulate these means of communication to fool us into believing what they want us to believe—a tool also used by politicians globally for their smear campaigns and fake news generation.

In this age of high-quality hyperconnectivity, creators, consumers and businesses need the best tools for the job to do what they do best. However, with tech companies like Apple releasing new devices with minor upgrades year upon year, and purposefully slowing down[73] the functionality of their previous editions, the consumer is left with no option but to make yet another expensive purchase and discard their old devices to the earth.

The digital age, to me, is a force for good—after all, how many people are fortunate enough to truly see the world? How many people can afford to learn from the wisest minds of our generation? How many people can gather a crowd large enough to hear their pleas for help with only their voices? The digital age has a widespread influence that positively impacts the lives of its consumers in multitudes of ways.

However, the influence of the digital age can also have damning effects. We must be cognizant of the fallout of our quest for convenience and entertainment, and do what we can to help where help is needed.

Life as an influencer is not easy—being online all the time comes with its own sense of responsibility and obligation. This is why I feel it is my responsibility to do what I can to make my work mean something while causing as little harm as possible, and spreading a message of goodwill toward all.

But how can we, as consumers, make a difference?

HOW TO REDUCE YOUR INTERNET CARBON FOOTPRINT:

WATCH YOUR VIDEO STREAMING

• Turn off auto-play.[1]
• Close tabs that you are not using to prevent videos playing in the background.[2]
• Avoid using videos when you need audio.[3]

CHANGE YOUR EMAIL HABITS

• Limit "reply all".[1]
• Unsubscribe from newsletters you're no longer interested in.
• Talk in person rather than over email.

CHOOSE A CONSCIOUS CLOUD

• Consider storing your data on a green cloud provider.
• Some providers are run completely on renewable energy sources.

SHUT DOWN

• The U.S Department of Energy recommends powering down computers if you'll be away from them for more than 2 hours.

KILL THE VAMPIRE POWER

• When plugged in but powered down, computers continue to draw energy, this is known as vampire power.
• 1/4 of all residential energy consumption is used on devices in idle power mode.[2]

TAKE A SNOOZE

• Even in sleep mode, computers continue to burn energy.
• The average laptop burns 50-100 W/hour of electricity while in use and 1/3 of that in stand-by mode.[4]
• Set your computer to go into sleep mode after certain amount of minutes so it can run more efficiently while you're away for shorter periods of time.

DIM YOUR MONITOR

• Dimming from 100% to 70% can save up to 20% of the energy the monitor uses. Plus, lowering brightness reduces eye strain.[1]

USE A TABLET OR SMARTPHONE (SOMETIMES)

• For quick searches and non work-related tasks, use a tablet or smartphone instead of a laptop or desktop.[1]

OTHER TIPS

Hold onto your IT equipment for as long as possible. Get it repaired before you buy a new device.[4]

Unsubscribe from irrelevant newsletters.[3]

Go direct to the website rather than using a search engine. Save websites you visit regularly in favourites.[4]

Figure 6-14: The various methods of reducing your carbon footprint.[74]
Source: WebFX

Solutions

As a digital creator, I am all too aware of the harmful impacts of the digital industry. Here are a few means I have discovered to make my consumption patterns more sustainable, which you too can try.

Stream Less

While on-demand viewing and listening are among the biggest perks of living in the twenty-first century, it is also not something we cannot live without.

I for one do not have a subscription to any paid streaming service. Whether it's music, TV shows, movies or otherwise, I spend my time consuming free-to-watch YouTube content and not much else.

The most efficient and environmentally friendly way to consume content is to wait for it to come on terrestrial TV rather than viewing via video, and use an audio streaming platform that functions over Wi-Fi. Moreover, the carbon emissions of streaming video, games and music have never been higher and more energy-intensive across history.

Use Wi-Fi over Cellular Data

As anyone with a phone will tell you, there is nothing more frustrating than low network when you want to connect to the internet.

However, even though cellular networks offer continuous connectivity over a larger area compared to Wi-Fi, they involve the use of terrestrial transmitters to access the internet, in contrast to router-based transmitters for Wi-Fi. This causes the device to consume more energy thereby making it less efficient.

Using a phone with Wi-Fi is half as energy-intensive as compared to a phone with a mobile network.[75]

Decarbonize Storage and Search

Choosing clouds whose data centres rely on renewable energy can minimize carbon emissions while being more energy efficient.

Using cloud servers means companies produce half the amount of carbon emissions they otherwise would.[76] Google, Amazon and Microsoft are some of the top sustainable and green cloud providers, according to reports published by the companies themselves. These claims are based on the fact that these companies, while not necessarily producing renewable energy themselves, purchase it from other sources for use in their manufacturing and operations.

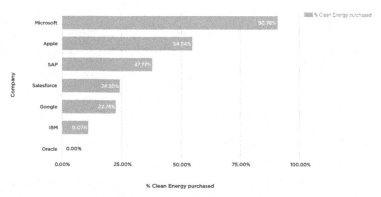

Figure 6-15: Clean energy purchased vs company.[77] Source: Green Monk

Vendor	Q4 2018 (US$ billion)	Q4 2018 Market share	Q4 2017 (US$ billion)	Q4 2017 Market share	Annual growth
AWS	7.3	32.3%	5.0	32.2%	+46.3%
Microsoft Azure	3.7	16.5%	2.1	13.7%	+75.9%
Google Cloud	2.2	9.5%	1.2	7.6%	+87.7%
Alibaba Cloud	1.0	4.2%	0.6	3.5%	+73.8%
IBM Cloud	0.8	3.6%	0.6	4.2%	+27.6%
Others	7.7	33.8%	6.1	38.9%	+26.7%
Total	22.7	100.0%	15.6	100.0%	+45.6%

Table 6-3: Worldwide cloud infrastructure spending and annual growth.[78]
Source: Canalys

Ecosia, a sustainable search engine launched on 7 December 2009, marking the UN Climate Talks in Copenhagen, offers to plant a single tree for every forty-five searches via its engine in an attempt to offset its carbon footprint and enable the reduction of carbon from the atmosphere.[79] However, the success of such projects depends on how long the trees take to grow to maturity, and what happens after.

Ecoasia and other such companies represent a ray of hope for the future, where conscious decisions by service providers ease the environmental obligation on its consumers.

Limit Screen Time

While we do indeed live in the age of technology, it is not impossible to limit our use of our devices. When you are not working or studying, try and limit your screen time as much as possible. Most smartphones and tablets today have customizable systems in place for you to audit and limit the amount of time you spend on your devices, even including a breakdown of your activity.

Limiting your screen time not only limits your use of the internet, but it allows time for your eyes to rest, your mind to disengage from the constant barrage of information, and for you to instead find ways to entertain yourself in the natural world.

Green Your Inbox

While emails are not the most harmful form of communication, the volume and storage associated with global use can be the difference between widespread destruction and a global shift towards sustainable internet use.

Here are a few ways you can make your inbox greener:

- Delete emails that you won't need again to prevent them from being stored unnecessarily.
- Unsubscribe from email newsletters and mailing lists that you never read.
- Avoid gratuitous thank you emails.
- Use a shared server for heavy files, rather than storing them on your emails.

Prolong Your Gadget's Life

In the modern world, there is no lack of new devices hitting the market every other day. In today's consumer-forward world, it can be tempting to upgrade as often as possible, to keep up with the latest trends and features.

However, this use-and-throw mentality results in the devices not being used to their full potential, making them highly ineffective from both an energy and a monetary perspective. This thereby increases our carbon footprint as well as the digital footprint of the devices themselves.

If we're able to use our devices longer, we can reduce the number of them produced, which will save vital limited resources and conserve much-needed energy.

Here are some tips on how to prolong the life of your gadget:

- **Avoid overcharging**: Ever noticed your device seems to run out of battery quicker than it used to? Chances are, you are probably over or undercharging your device. Overcharging is one of the most common reasons for issues with battery life longevity. Most manufacturers recommend not leaving your gadget charging overnight—instead, charge your device only until the battery is full. On the other hand, letting your battery drain entirely is also not considered a good practice. As a rule, try and stay between 15 per cent to 95 per cent charge.
- **Cleanliness is key**: While keeping your devices clean is a no-brainer, sometimes, without us even noticing, fine particles of dirt and fibres can make their way into your devices, greatly affecting the longevity of your devices. This

can cause overheating, among a host of other issues that can adversely affect the longevity of your device.

- **Stay up to date**: Updating your software and operating system is crucial to the functioning of your device. With a host of bug fixes, stability improvements, security optimizations and more, updating your device regularly is important to its overall life cycle.
- **Keep them dry**: Water is the worst enemy of your devices. While we should always avoid spilling, water can enter your devices through rain as well as moisture in the air. As far as possible, try not to take your devices into the bathroom with you while you shower, and if you live in a humid place, opt for a dehumidifier or a weather-sealed box for devices you do not use regularly. There are also a host of waterproofing accessories available on the market, so do your research and pick one that is suitable to your needs.

One of the best ways to extend the life cycle of your products is to buy smarter, more sustainable products.

Here is a simple list of questions to help you make smarter choices:

- Is it relatively easy and inexpensive to repair?
- Can the battery be replaced?
- If disposal is the only option, does the manufacturer provide options for exchanges or proper disposal?

Next time you are considering upgrading, ask yourself whether you are doing so out of necessity or for style.

Dispose of Your Devices Properly

E-waste is a global problem, and unlike unethical manufacturing processes, can be mitigated by the consumers themselves. For example, applications like Cashify[80] provide a platform that helps people sell and upcycle electronic devices and their accessories by refurbishing, repairing and recycling them.

Here are a few ways in which you can dispose of your devices in an eco-friendly way:

- **Recycling:** Recycling devices is one of the easiest and most efficient ways to dispose of your devices responsibly. The best part? The original retailers can do this for you. Under the Extended Producer Responsibility (EPR) regulations for E-Waste,[81] most electronic companies employ Producer Responsibility Organizations (PROs) such as Karo Sambhav,[82] rPlanet,[83] and Namo E-Waste,[84] among others, and are obligated to recycle products sold to consumers. Several brands have collection boxes through which one can return used devices.
- **Repair:** Instead of investing in a new device, check with your closest retailer or IT store to see if your device can be repaired. This keeps used devices out of landfills and extends their life.

In many cases, developers and manufacturers today are opting for greener solutions for their consumers. One recent example of this is Apple's new iPhone 14. According to their Product Environmental Report,[85] which details the sustainable steps they have taken to minimize e-waste, maximize raw material use and avoid unnecessary use of harmful and toxic chemicals, the report details their plans from sourcing to disposal—providing users a level of transparency and choice that has so far been missing from the Apple brand.

Mandeep Manocha, founder and CEO of Cashify, says:

'Being "Responsible" costs. Communicating, collecting and recycling costs are high and rarely covered by the revenue generated by sales of recovered materials. Brands do not generally wish to bear these costs and would rather pass them on to customers. Cashify extends the life cycle of digital products by refurbishing them and making them available to people at lowered costs. This prevents old unused devices lying in people's drawers from turning into e-waste. In effect, we are delaying e-waste generation.'

Use Your Devices Responsibly

While proper disposal is important, we can start making more conscious decisions for our digital footprint by making small changes to our habits at this very moment.

Here are a few things you can do with the devices you're using currently:

1. Dim your monitor.
2. Employ battery saver mode.
3. Do not use your devices while charging them— this reduces your battery life.
4. For quick searches, avoid using your desktop or laptop.
5. Visit a website directly rather than using a search engine.
6. Delete unused apps, redundant screenshots and photos from iCloud or other cloud drives.

If someone were to ask me which was the most sustainable phone on the market, my reply would be, 'The one in your hand.'

The device you are currently using is associated with its own carbon footprint and every day you use that phone is a step closer to making up for it. However, if you choose to buy another phone before the phone you are using has reached the end of its life, no matter how sustainable the other phone is in terms of its raw materials, manufacturing process, energy requirements or even its recyclability, its carbon footprint is still bigger than that of the one in your hand.

If you are looking to upgrade to a new device, look for one that suits your needs, and then try and find out its carbon footprint. Your goal as a consumer should always be to minimize the damage caused by the carbon footprint of your device by using it to its most optimal degree.

For example, my iPhone 14's life cycle accounts for 61 kg of carbon emissions. My goal is to use the phone for a minimum of four years, as I do with all my devices, offsetting its unsustainability.

While the age of hyper-connectedness is both a boon and a curse, it is up to us, the consumers, to change the way we make use of this gift of technology and innovation.

Resources

Here is an inventory of supplementary media material, people who matter and shops in line with the chapter's theme.

Books

1. *Speed and Scale: An Action Plan for Solving Our Climate Crisis Now*
    Author: John Doerr
    Published by: Penguin Books
    https://speedandscale.com

2. *Digital Minimalism: Choosing a Focused Life in a Noisy World*
    Author: Cal Newport
    Published by: Portfolio
    https://www.calnewport.com/books/digital-minimalism

3. *Drawdown: The Most Comprehensive Plan Ever Proposed to Reverse Global Warming*
    Author: Paul Hawken
    Published by: Penguin Books
    https://drawdown.org/the-book

4. *Fossil Free: Reimagining Clean Energy in a Carbon-Constrained World*
    Author: Sumant Sinha
    Published by: Harper Business

https://sumantsinha.com/book-fossil-free-reimagining-clean-energy-in-a-carbon-constrained-world-authored-by-sumant-sinha/

Documentaries

1. *Planet of the Humans*
    Directed by: Jeff Gibbs
    Available on Amazon Prime, Youtube Movies
    https://planetofthehumans.com/

2. *An Inconvenient Truth*
    Directed by: Davis Guggenheim
    Available on Amazon Prime
    https://www.amazon.com/Inconvenient-Truth-Al-Gore/dp/B000KZ3BWE

3. *Tomorrow*
    Directed by: Cyril Dion and Melanie Laurent
    Available on Amazon Prime
    https://www.tomorrow-documentary.com/

4. *The Third Industrial Revolution: A Radical New Sharing Economy*
    Directed by: Eddy Moretti
    Available on YouTube
    https://www.vice.com/en/article/bj5zaq/watch-vices-new-documentary-the-third-industrial-revolution-a-radical-new-sharing-economy

5. *The Social Dilemma*
    Directed by: Jeff Orlowski

Available on Netflix
https://www.thesocialdilemma.com/

Ted Talk

1. How India Could Pull Off the World's Most Ambitious Energy Transition
   By: Varun Sivaram
   https://www.youtube.com/watch?v=Pgq_CODucg0

2. How Solar Power Can Help India Become a Super-Power
   By: Kunal Munshi
   https://www.youtube.com/watch?v=FdG21aapiCY

3. The 'Greenhouse-in-a-Box' Empowering Farmers in India
   By: Sathya Raghu Mokkapati
   https://www.youtube.com/watch?v=Anlyzh RX9IM&list=PLOGi5-fAu8bHbbepxw2rp2ll29SC1D DH_&index=4

4. Smart Green World? Making Digitalization Work for Sustainability
   By: Tilmann Santarius
   https://www.youtube.com/watch?v=lNkaGLMlm_Q

5. Batteries Not Included
   By: Marek Kubik
   https://www.youtube.com/watch?v=ffG7qW2l024

Shops

1. Tata Power Solar
    Solar Panels for Homes
    https://www.tatapowersolar.com

2. Zolopik E-Waste Recycling
    Sell Your E-Waste
    https://www.zolopik.co

3. Namo eWaste
    E-waste recycling
    https://namoewaste.com/

4. Air-Ink - Ink from Air Pollution
    https://air-ink.com

5. Phillips
    Smart Lighting for Homes
    https://www.lighting.philips.co.in/welcome

6. Ecobee
    Save Energy With a Thermostat
    https://www.amazon.in/ecobee-Smart-Thermostat-Enhanced-Compatible/dp/B09XXTQPXC

Tools

1. Freedom
    Block websites, apps, and the Internet
    https://freedom.to/

2. New feed eradicator | Chrome extension
    Eradicate social media noise
    https://chrome.google.com/webstore/detail/
    news-feed-eradicator/fjcldmjmjhkklehbaci
    haiopjklihlgg

7

Home

GHG Impact: 36 per cent[1]

This industry impacts the following Sustainable Development Goals (SDGs):
- SDG 7: Affordable and Clean Energy
- SDG 9: Industrial Innovation and Infrastructure
- SDG 11: Sustainable Cities and Communities
- SDG 15: Life on Land

ACTIVITY

Before we begin this chapter, try to answer the following questions as honestly as possible:

1. Where do you live?
    - ☐ Multi-family residential building
    - ☐ Apartment complex
    - ☐ Single-family residential building

2. Do you have a home composting unit?
    - ☐ Yes
    - ☐ No

3. Do you have a terrace farm/herb garden?
    - ☐ Yes
    - ☐ No

4. Do you use any of the following ways to save water at home?
    - ☐ Flow restrictors
    - ☐ Rainwater harvesting
    - ☐ Reusing greywaters
    - ☐ Others
    - ☐ None of the above

5. Do you use any of the following devices to save energy at home?
   ☐ Thermostat
   ☐ Solar powered devices
   ☐ Smart Led Lights
   ☐ Automatic device controller
   ☐ Others
   ☐ None of the above

Since the age of the cave-dwellers, humankind has longed for a place to call their own. The foundation stone of Maslow's Hierarchy[2] [3] of Needs and a core feature of the internationally recognized mandate on Fundamental Human Rights, shelter has played a critical role in humanity's journey to civilization.

As we moved from caves to mud homes to brick structures and high-rises, this fundamental need transformed. From the Royal Buckingham Palace to Trump Towers, and from Mumbai's controversy riddled Antilia to The World Islands in Dubai, our homes and structures mirror our personalities, beliefs and social standing.

Today, our homes are not just where we live, but are reflections of *how* we live.

Where once construction activities were limited to housing and dictated by their proximity to natural resources and strategic advantages, such as rivers, arable land and neighbouring settlements, today, high-speed motorways, towering statues, sprawling housing societies and national monuments constitute most large-scale infrastructure projects.

Today, we build not only out of necessity but out of an unstoppable need to create. All around our nation, every day, homes are built on the strength of countless backs and the calloused skin of bare hands.

The global infrastructure construction market was valued at USD 2,242.3 billion in 2021 and is expected to reach USD 3,267.3 billion by 2027.[4] The construction sector is also the second-largest employer in India, with over 51 million people employed across the country.[5]

I have always had an unusual relationship with home. To me, home is not a place, but a feeling. Home is where I rest my head for the night, wherever in the world that may be. Over the years, I have had many homes, in just as many places, with no two ever being the same.

I have seen the tallest skyscrapers and climbed taller mountains. I have marvelled at the architectural feats of man and gazed in awe at the boundless wonders of nature. As humans, we have always tried to replicate the raw power of nature—the ability to create, destroy, nurture and annihilate as we see fit.

But in this quest for power, convenience, status and luxury, what have we left behind?

To Build a Home:
The Problems with Infrastructure

To unravel the varying problems with infrastructure and the way we live, we would have to tackle the issue one step at a time, beginning with the very first building block—building materials.

Global consumption of building materials tripled from 6.7 billion tonnes in 2000 to 17.5 billion tonnes in 2017, with concrete, aggregates and bricks being the most commonly used building materials. However, not all materials are made equal.[6]

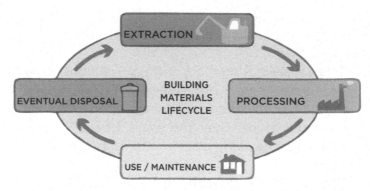

Figure 7-1: Building materials lifecycle.[7] Source: Mizan-Tepi University

Building materials globally consume 30-50 per cent of available raw resources and produce about 40 per cent of waste in landfills in Organisation for Economic Co-operation and Development (OECD) countries. For example, materials such as wood come from natural sources, with millions of trees being felled for use as raw materials. Some of these materials, such as the highly sought after teak wood, are priced expensively due to the felling of teak trees, which take twenty to twenty-five years to grow to maturity and provide some of the most durable and attractive wood on the market.

Of all building materials, cement emerges to be the most problematic from an ecological perspective. Cement is one of the most widely used materials for construction, being the basis of buildings, bridges, monuments and more. According to some, it is also the highest consumed product on earth besides water.

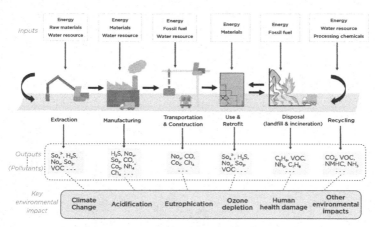

Figure 7-2: Different emissions and their environmental impacts.[8] *Source: Cell Press Open Access*

This is a harrowing fact, taking into account that the amount of CO_2 emitted by the cement industry is nearly 900 kg for every 1000 kg of cement produced.[9]

Cement production makes use of lime, a raw material that must be mined and quarried, disturbing local biodiversity and contributing to air pollution through the release of vast quantities of dust and other particulate matter.

Cement manufacturing contributes to GHG emissions both directly and indirectly, through the heating of calcium carbonate, which produces lime and CO_2, and via energy use and the associated burning of fossil fuels in factories. The process also releases nitrogen and sulphur oxides as well as carbon dioxide and carbon monoxide emissions.

These emissions are responsible for global warming, ozone depletion, acid rain and biodiversity loss, leading to a host of other issues such as reduced

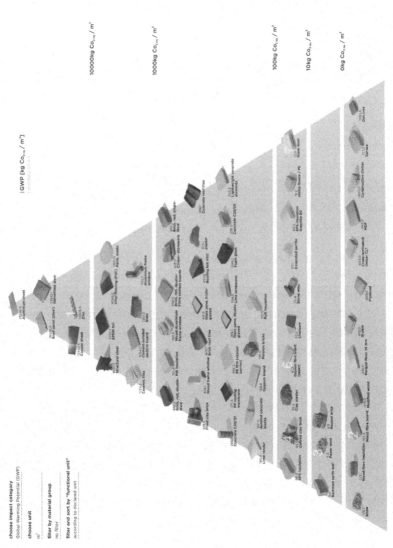

Figure 7-3: Environmental impact of different building materials.[10] Source: Arch Daily

crop productivity and more. Moreover, these harmful emissions can also affect human health, with reports of workers facing ocular discomfort from sediment settling in their eyes, respiratory diseases like asthma, bronchitis, tuberculosis, cardiovascular diseases and in extreme cases, even death.

As much as 50 per cent of whole-life carbon emissions in a building can be attributed to embodied carbon, i.e., carbon emitted via the manufacturing of building materials and the process of construction itself, the majority of this figure being produced at the start of the structure's life cycle. As few as six materials account for over 70 per cent of the construction-related embodied carbon.[12]

The building materials we use affect the integrity of the eventual structure, which is why the combinations

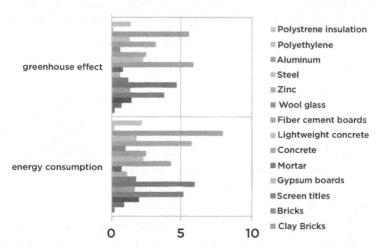

Figure 7-4: Emissions attributed to embodied carbon[11]. Source: Research Gate

of materials used across modern history have remained largely the same.

However, the most commonly used materials, due to the frequency of their use, are highly unsustainable for the environment.

Let's look at some of these materials through the lens of their impact on the environment.

Aluminium

A lightweight and malleable material, aluminium is a common feature around construction sites. A great conductor of heat, it is the material of choice for most heating and electrical features.

Aluminium in its raw form, however, is not fit for immediate use. Raw aluminium ore must first be mined in a highly energy-intensive undertaking, and then processed to be commercially sold and used. This process is so highly unsustainable that the production of aluminium alone is responsible for approximately 2 per cent of global GHG emissions, i.e., 1.1 billion metric tonnes of CO_2 every year.[13]

Wood

One of the oldest building materials, wood has been a constant feature in manmade structures since the early age of man. Versatile and aesthetic, wood has long been touted as a renewable resource, but due to its overuse, is causing more harm than good to our environment.

While wood is sourced from trees, meaning that the more trees we grow, the more wood we have available, the growth and maturity cycle of the trees themselves is a long process and cannot keep pace with global consumption. We are cutting down trees faster than they can be grown. As living, breathing carbon stores and recyclers, the felling of trees on such a large scale causes global spikes in CO_2 emissions.

Asbestos

A mineral fibre commonly used for insulation and fire-safety proofing, asbestos is a strong and durable fibrous mass that is ideal for construction activities. However, asbestos can enter our bodies via inhalation, leading to potentially life-threatening consequences.

While asbestos has been banned in many countries around the world, including India, the highly carcinogenic component is still a major feature in construction activities in developing nations.

Steel

One of the most widely used materials in construction, steel is one of the most versatile and useful materials known to man. However, its production is associated with an equally significant carbon footprint.

The process of manufacturing steel is both water and energy-intensive and emits GHGs in vast quantities. Every tonne of steel produces an average of 1.85 tonnes of carbon dioxide, accounting for about 8 per cent of global carbon dioxide emissions.[14]

Glass

From windows to doors and installations, glass is an aesthetically pleasing and valuable material. Made using sand, a non-renewable resource, glass fixtures have become increasingly popular for their delicate and reflective nature.

However, glass production is highly energy and water-intensive and is a key contributor to global GHG emissions, with 86 metric tonnes of CO_2, or 0.3 per cent of global emissions, produced every year.[15]

Lead

Lead is a heavy metal known for its durability and malleability associated with its low melting point. Often used in pipes and plumbing, lead is a popular building material.

However, as most renaissance art enthusiasts and art historians will tell you, it is also an extremely toxic material that can lead to severe health hazards, and even death, if accidentally ingested or inhaled.

Lead can also pollute the air, water and soil, with the dust remaining in the air indefinitely, leaching into the water and the soil, affecting all living things. Lead is also known to hinder plant growth by affecting its natural photosynthesis processes.

While lead is not as widely used as it once was, it still prevails in unregulated construction activities in developing nations.

Power Struggle: Energy and Water Consumption

When speaking of the way we live, it is hard to separate our daily lives from the energy we consume to carry out the numerous activities we undertake.

From the lights we use to brighten the indoors to temperature control systems like heaters and air conditioners, and from our many entertainments and electronic devices to elevators and electronic locks, energy is utilized in homes and buildings in massive quantities every day.

According to a UN report that detailed energy use in buildings across Asia, buildings' share of energy use in 2019 ranged from 49 per cent in China to 25 per cent in India, and 23 per cent in the Association of Southeast Asian Nations (ASEAN) region. In buildings in China, electricity accounted for 35 per cent and biomass for 15 per cent of energy use, while in India electricity accounted for 19 per cent and biomass for 55 per cent.[16]

In developing countries like India, the annual growth rate in air conditioners has been the most prominent, at 15.5 per cent every year.[17] This coincides with a rise in heatwaves and heat-related deaths associated with global warming, disproportionately affecting southeast Asian regions, where average temperatures have been on the increase every decade since 1960.[18]

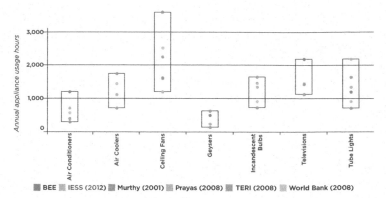

Figure 7-5: Annual appliance usage hours.[19] Source: Research Gate

Another important metric to keep in mind while discussing the way we live is our consumption of water. While a person requires 4 litres of water a day

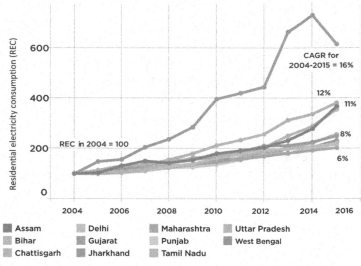

Figure 7-6: Residential electricity consumption.[20] Source: Centre for Policy Research

to stay hydrated and function with ease in moderate climatic conditions, on average, we consume around 4 trillion cubic meters of fresh water globally in a year.[21]

Water is a non-renewable resource, and with a rapid increase in droughts and changing rainfall conditions, water shortages have been a cause for global concern. In developing nations lacking appropriate sanitation measures, wastewater is dumped directly into large water bodies, causing mass eutrophication and limiting access to clean, consumable water.

In 2014, India had the largest freshwater withdrawals at over 760 billion cubic metres per year. This was followed by China at just over 600 billion metric tonnes and the United States at around 480-490 billion metric tonnes.[22]

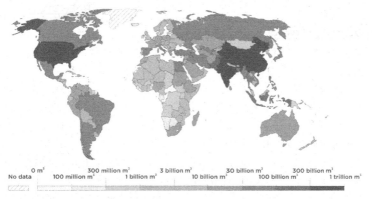

Figure 7-7: Freshwater withdrawal in 2017.[23] Source: FAO

Gone to Waste: Household Waste Management

How much waste do you produce in a day? From plastic wrappers for products to food scraps, garbage and even sewage, the waste generated per person per day across the world averages 0.74 kilograms, ranging from 0.11 to 4.54 kilograms depending on one's socio-economic background and country of origin.

Every year, the world generates 2.01 billion tonnes of municipal solid waste, at least 33 per cent of which is not managed in an environmentally safe manner.[24, 25]

In the Indian context, urban settlements produce approximately 42.0 million tonnes of municipal solid waste annually, i.e., 1.15 lakh metric tonnes per day. Of this, 83,378 tonnes are generated every day in 423 Class-I cities. Municipal solid waste comprises 30 per cent to 55 per cent biodegradable or organic matter,

APPENDIX-A

Quantity of waste generation

Total quantity of municipal solid waste generated - 1.15 lakh tonne / day (TPD)

		% of total garbage
Waste generated in 6 mega cities	21,100 tpd	18.35%
Waste generated in metro cities (Population 10 lakhs +)	19,643 tpd	17.08%
Waste generated in other Class-I towns (1.0 lakh plus population)	42,635.28 tpd	37.07%
	83,378.28 tpd	**72.50%**

If waste produced in all 423 class-I cities is tackled, percentage of solid waste managed scientifically would be 72.5% of total waste generated each day.

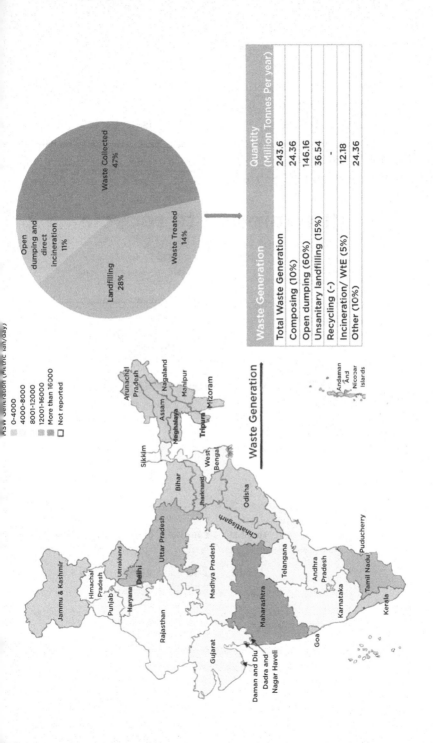

MSW generation (Metric Tah/day)

- 0-4000
- 4000-8000
- 8001-12000
- 12001-16000
- More than 16000
- Not reported

Waste Collected 47%

Open dumping and direct incineration 11%

Landfilling 28%

Waste Treated 14%

Waste Generation

Waste Generation	Quantity (Million Tonnes Per year)
Total Waste Generation	243.6
Composing (10%)	24.36
Open dumping (60%)	146.16
Unsanitary landfilling (15%)	36.54
Recycling (-)	-
Incineration/ WtE (5%)	12.18
Other (10%)	24.36

Jammu & Kashmir

Himachal Pradesh

Punjab

Uttrakhand

Haryana

Delhi

Rajasthan

Uttar Pradesh

Gujarat

Daman and Diu
Dadra and Nagar Haveli

Madhya Pradesh

Maharashtra

Goa

Karnataka

Telangana

Andhra Pradesh

Chhattisgarh

Sikkim

Bihar

Jharkhand

West Bengal

Odisha

Arunachal Pradesh

Assam Nagaland

Meghalaya Manipur

Tripura Mizoram

Kerala

Tamil Nadu

Puducherry

Andaman And Nicobar Islands

40 per cent to 55 per cent inert matter and 5 per cent to 15 per cent recyclable waste.[26, 27]

Just because the waste is recyclable or biodegradable doesn't mean that it is handled sustainably. Recycling and segregation of waste, if not undertaken at the source, is a tedious and taxing

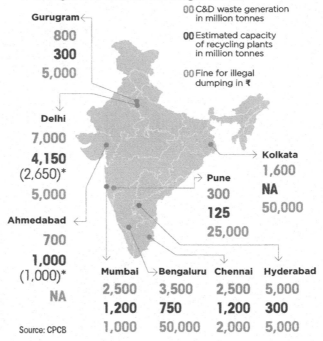

Ill-equipped to recycle

Plants do not have the capacity to manage the waste cities generate

00	C&D waste generation in million tonnes	
00	Estimated capacity of recycling plants in million tonnes	
00	Fine for illegal dumping in ₹	

Gurugram
800
300
5,000

Delhi
7,000
4,150
(2,650)*
5,000

Kolkata
1,600
NA
50,000

Pune
300
125
25,000

Ahmedabad
700
1,000
(1,000)*
NA

Source: CPCB

Mumbai	Bengaluru	Chennai	Hyderabad
2,500	3,500	2,500	5,000
1,200	**750**	**1,200**	**300**
1,000	50,000	2,000	5,000

*Additional capacity of plants in the pipeline.
All figures are based on the capacity of the prposed recycling plants except for Delhi and Ahmedabad, which have already set up their plants.

Figure 7-8: Plants do not have the capacity to manage waste generated by Indian cities.[28] Source: Central Pollution Control Board

process, with many developing nations ill-equipped to take on the burden.

Instead, this waste is dumped into landfills or waterbodies, polluting the soil, air and water and affecting the natural integrity of our fragile ecosystems.

Growing Pains—Over-population and City Planning

Now that we have discussed some of the ways in which our homes are unsustainable, let's zoom out and examine the problem on a larger scale, and from the perspective of the consumer—the populace.

All around the world, more humans are living in both urban and rural settlements than ever before. According to a UN report, In 1950, the world population was estimated at around 2.6 billion people, reaching 5 billion in 1987 and 6 billion in 1999.

In October 2011, the global population was estimated to be 7 billion. Today, the world's population is expected to increase by 2 billion persons in the next thirty years, from 7.7 billion currently to 9.7 billion in 2050, and could peak at nearly 11 billion around 2100.[29]

However, our planet's natural resources are limited, and cannot withstand the growing pressures of our demands. The issue is especially prevalent in nations like India, where rural-urban migrations have seen cities overflowing with people in search of viable means of income, and very few resources to afford even their most basic needs, such as food, water, shelter and clothing. For example, Mumbai, the commercial hub of India, is known as one of the most densely populated cities in the world, with approximately 73,000 people per square mile.[30]

Mumbai Population 2022
20,961,472

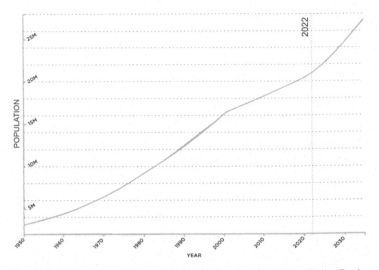

Figure 7-9: Rise in population of Mumbai.[31] Source: World Population Review

As per a NITI Aayog report[32], half of India will become urbanized over the coming years, with urban populations contributing to 73 per cent of the total population increase by 2036.[33]

However, the true problem with this growth in population isn't just the people, but the lack of planning associated with urbanization, with the environment, the economy and marginalized communities bearing the brunt of the adverse effects.

Aliste Technologies founder Aakarsh Nayyar says:

'The biggest source of household consumption is from Electrical appliance operation. Of this, 20–30 per cent electricity is wasted. Energy consumption patterns are not equitably distributed among different income groups as there is an exponential increase in energy consumption as we move towards the upper-middle and upper-income groups. This is due to a significant increase in the number of appliances and unconscious energy consumption. Thereby energy tracking and consumption accountability are vital among the higher-income groups. Currently, accessibility in remote areas and affordability for middle and lower-income groups are the major hurdles to the adoption of green energy in India.'

This lack of planning is evident in cities across the nation, which remain plagued by traffic; pressure on basic infrastructure; extreme air, water and noise pollution; urban flooding; water scarcity and droughts.

The infrastructure sector is a key growth driver for the Indian economy. A nation on the brink of mass development and complete globalization, India is investing in upgrading to international standards of infrastructural capabilities in a shorter period than ever before seen. However, this rapid development comes with a host of dire consequences for the natural world.

Let's examine how this rapid pace of playing catch-up with developed countries affects not only our local populations but also the natural world that we rely on for our sustenance.

Unacknowledged Urbanization

Due to a lack of a comprehensive census, a large part of urbanization in India is left unacknowledged and unaccounted for. Almost half of the 7,933 urban settlements are census towns—urban areas with a population of at least 5000, a density of at least 400 people per square kilometre, and at least 75 per cent of the male workforce engaged in non-agricultural labour. These census towns, while being urban centres, strictly speaking, are governed as rural entities through the panchayat system. As a result, they are often deprived of adequate urban planning.[34]

Haphazard Planning

City planning is an important matter when it comes to development. From the smooth flow of traffic to the optimal use of resources and infrastructure, security and even the potential for future development projects, all rely on an effective master plan for how a settlement, town or city is arranged. These plans guide and regulate city development and are critical to urban planning.

While this is true, 65 per cent of the 7933 urban settlements in India, i.e., approximately 5156 areas, have been built and function to this day without a master plan. This results in many issues, including a lack of pollution control measures, illegal and unplanned construction of slum housing and traffic jams, which are hazardous for residents and the environment.[35]

Moreover, once a city develops haphazardly, corrective measures come with much difficulty and expense, hindering the development of necessary infrastructure.

Poor Utilization of Land

Where there are people, there must be land on which they may settle. However, due to the population density and lack of proper city planning, most Indian cities are victims of poor land use. This sub-par utilization of land means that small sections of the land become densely populated to accommodate the city's growing population, as witnessed by the multitudes of slums in urban settlements like Mumbai.[36]

Build It and They Will Come: Large-Scale Infrastructural Projects

Did you know that every attraction on the famed Wonders of the World list involves varying degrees of man-made structures?

From the marble façade of the Taj Mahal to the ancient Inca dry stone walls of Machu Picchu, humans have left their mark on our planet with our ingenuity and creativity. Famously, the Great Wall of China can even be seen from space!

If you thought these iconic structures were a thing of the past, think again.

Recent years have seen countries on a mad dash to create the next tallest building, the next Statue of Liberty, or any other national symbol that attracts tourists and their foreign currency to their shores. But what does it take to create the next enduring symbol of humanity's existence?

Famously, the Indian government recently sanctioned and built the Statue of Unity, which was unveiled in 2018 and formally recognized as the tallest in the world. While the project was declared a success, its construction spelt doom for the area's ecological balance and displaced indigenous communities which previously resided there.[37]

Large-scale development projects affect not just the people living in surrounding areas, but the sanctity, safety and integrity of the water, soil, vegetation and wildlife too, as demonstrated by this graph.

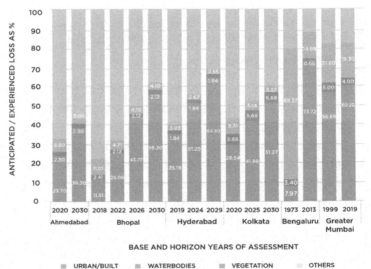

Figure 7-10: Vegetation loss and urban development over the years.[38]
Source: Observer Research Foundation

While roads, highways, airports, and more are necessary to the overall development of a region, monuments and structures, such as the Statue of Unity, are examples of poor land use, and in a nation as large and population-dense as India, this can be the difference between shelter and mass homelessness.

Life in the Big City: Pollution and Health Hazards

Climate change and the pollution associated with life in busy metropolises can have serious impacts on our health as well as that of the environment. With much of the country still undergoing a move from rural to urban life, sustainable development must be emphasized to avoid large-scale misuse of resources, and the resulting air, water, land and noise pollution.

This trend of urbanization is only projected to pick up steam over the coming years, with sharp increases in anthropogenic emissions to match. The effects of these changes are not alien to those living in developing nations—from extreme precipitation that causes flooding to droughts that lay waste to entire harvests, and from higher temperatures to frequent heat waves, typhoons, cyclones and more.

Adding to the impacts on the local climate, the rapidly increasing urban population also means higher numbers of vehicular traffic and congestion, in turn increasing pollutant emissions and aerosol load in the atmosphere. The increasing urbanization, growing population and industrialization has been stated as one of the key reasons for high aerosol loading in the Indian sub-continent.[39]

To understand why this is a problem, let's take a closer look at aerosols, and unravel what they mean for our environment and our health.

Aerosols

Defined as suspended particles of either a solid or liquid nature, aerosols can be both naturally occurring or can be formed as the result of human activities.[40]

These microscopic particles can be formed in several different ways, from windblown dust and volcanic activity to salty sea spray, as well as fossil fuel combustion in automobiles, aeroplanes, ships and factories.

While some aerosols are relatively harmless in small quantities, carbonaceous aerosol sprays can contain highly toxic chemicals, neurotoxins and carcinogens that are extremely hazardous for adults, children and even domestic animals.

According to a report by WHO, more than seven million people across the world lose their lives due to diseases linked with pollution and particulate matter. India, being a rapidly developing country with an increasing population, is suffering from severe air pollution; of the world's ten most polluted cities, nine lie in India. The increasing air pollution in most of the Indian megacities over the last few decades and its consequential human health impacts (such as asthma and cardio-respiratory illness) have drawn prominent attention in recent years.[41]

What Are We Missing?

The way we live has always been a conundrum to me. As a nomad, I don't see my home as just four walls that contain all my belongings, but rather the world itself, with its thousands of people and uncountable mysteries.

I believe that this is why I can see what so many else miss—that we are trapped in a self-fulfilling prophecy of environmental destruction, a vicious loop that ends only with the destruction of our planet.

Think about it like this—when it gets hot in the city, people resort to temperature-controlling equipment such as air conditioners, freezers and more to make themselves comfortable. As more and more people switch on, the emissions from their systems are released in vast quantities into the atmosphere. These harmful emissions catalyse our planet's natural greenhouse effect, trapping more heat in our atmosphere. As it gets hotter, people turn up their ACs.

The Cycle Continues

However, this is not the only way to live. Our use of electronic gadgets can be viewed as our need to control the natural world, whether it's the heat, the breeze, the sunlight illuminating your study desk or the temperature of the water in which we bathe—we are deliberately defying nature to get our way, even if it means making what is considered impossible possible through the means of large-scale destruction.

This Is Not the Way It Has Always Been

I have always believed that as a generation, we have much to learn from our elders and their ancestors. Parents and grandparents love telling tales of how they travelled kilometres on foot to reach their schools and gain a proper education to support their families or bathed in the freezing waters of nearby rivers, streams and oceans. They tell of days when they would rub dirt into their wounds, a natural antiseptic against all diseases known to their childhood selves, or how, in the winter months, the family would huddle together in a single room to share what little warmth their bodies could muster.

They tell these stories because these stories are true, and because, in those days, living was a matter of going with the flow of nature, not against it. These stories are also true in the modern day, and in the rural villages that I frequent, where living with the flow of nature is not a lifestyle choice, but a way of life.

Ancient civilizations like the Indus Valley settlement are still heralded as architectural marvels and are stunning examples of building *for* the land, rather than just *on* it. With a working sanitation system, sewage lines and access to water from the nearby Indus river and its tributaries, the civilization is said to have prospered and thrived for over 700 years—all without overburdening the land on which it was built.

This is not to say that people in the modern world live only unsustainably. In the villages around my current home in Ladakh, sustainable living is not a choice, but the difference between life and death, starvation and satiation, and finding shelter and freezing out in the open.

Ladakh is a region that exists at the mercy of nature, where extreme temperatures and living conditions are the norm. From stifling heat that zaps the energy from your body in mere seconds spent under the sun, to frigid winds that all but freeze your toes off your feet, the barren, arid and high-altitude climate of the region is a testament to nature's power, and how mankind's dominion over the natural world can only go as far as his ability to be humbled.

I spent the winter of 2021-22 in these frigid lands, where temperatures are known to drop to -25 degrees. To survive in these temperatures, houses in Spiti are built to self-insulate, with every facet of the structures following the ancestral wisdom of locals from ages past.

Homes in Spiti are built using mud, to keep cool in the summer months and maintain the heat when it gets cold. The outer walls are painted in thick, black paint, to absorb the heat coming from the sun. The houses themselves are structured to include a multitude of common areas, so that families may stay warm by relying only on their body heat. In the harsh winter months, livestock is moved indoors too, with their dung being used to stoke fires and keep warm. Open wood fire stoves are used to cook, with their heat being trapped within the mud walls. Each home is built with its little farm, where non-water-intensive crops like peas are grown to feed those living within.

Water scarcity in the region is also combated through environmentally conscious means. Spiti Ecosphere, a local NGO, is known for its sustainable development projects all across the region. A few years ago, I was fortunate enough to participate in the building of an artificial glacier with the Ecosphere team, through which to relieve some of the locals' dire need for water. By working with the land, the Spiti Ecosphere team sourced giant boulders and rocks from nearby mounds and built a dam on a stream originating from a glacier that was rapidly receding away from local settlements. When it gets cold, the water collected in these dams freezes over, allowing locals to harvest it as and when needed.

Another indigenous example of Spitian wisdom is their dry compost toilets. A land rife with water scarcity, locals in Spiti must also be conscious of their water use. With very little water to spare for drinking, personal hygiene and irrigation, these dry compost toilets are built by digging a hole in the ground, with a mixture of dung, ash, and wood chips being used to help the waste decompose. Once the toilet is filled, it is closed up, and another is opened while the water stored in the previous hole decomposes. This cycle is repeated every season, with little to no water being wasted.

Ladakh and Spiti are just one example of indigenous communities living in harmony with nature, but India is home to thousands more.

Take Auroville in Pondicherry, for example. From a bird's eye view, the city is a vision in white—this is not just for aesthetic reasons. With a humid, warm climate, homes in Auroville are painted white to reflect heat and have wide open floor plans to allow the passage of the breeze, keeping it cool indoors.

In the not-so-distant past, India was known as one of the most conscious populations in the world. This is because the multitudes of belief systems indigenous to our nation all speak of the value of respecting nature and living with the land rather than against it. In most Indian households, waste is considered an inauspicious matter, and the value of buying only what one can consume is still of the highest priority.

However, the last few decades have seen this ancestral wisdom lost to the tales of our grandparents and ancestors, with a new capitalistic mindset of status and privilege reigning supreme. Today, we gratuitously waste water, food and energy in the name of comfort, not realizing that it is our very lives that hang in the balance.

I know what you're thinking—I live in a city. I do not have access to clean natural resources like rivers and streams, and cannot walk kilometres to get to work every day. How could I possibly make the change myself?

To this I say, you are not wrong. The problem is not that people do not want to live in harmony with nature, but that we are often ill-equipped to do so.

Three years ago, I was involved in the planning of my friend's new home, which was to be built in a manner that would make it self-sustaining and ecologically sound. From self-insulating walls to rainwater harvesting mechanisms, solar panels and more, the blueprints of the house painted a picture of the perfect sustainable home. However, the home that stands there today is a far cry from the vision we all held so dearly in our hearts.

So What Went Wrong?

The experience of designing this home taught me many lessons, but the two most pertinent were ones that were deeply disappointing to me, personally. For one, building a truly sustainable home in the city is not a cost-effective undertaking. With a lack of readily available technologies on the market, sustainable living comes with a hefty price tag, making it sustainable for the environment, but not for the finances of the home-dwellers.

Secondly, there is a widespread lack of education and understanding regarding building and construction activities that make optimal use of the natural world. While our plans were foolproof, the workers involved with the actual building process could not always execute what needed to be done, leaving us with few options but to go with what was known and comfortable to them.

However, these problems are not set in stone and promise to change with even a small shift in consumer mindset. What we need to realize is that just because we live this way today, that doesn't mean we are doomed to unsustainable living systems for the rest of our lives. We can still flip the script on our societal structures, and cause a ripple effect that would change not just the way we live, but the way the world and all its inhabitants will live until the end of time.

By changing the way we live, we are changing the world. I truly believe that all our problems, whether it is homelessness, disease, hate or war, arise from our lack of alignment with the natural world. When we realign ourselves, we will restore the world to a state of equilibrium—a critical balance that must be striven for and protected at all costs.

Brick by Brick: Solutions

So how do we go back to living in harmony with nature? Do we have to resort to simple lives in small village-like settlements, where water must be harvested from lakes and rivers and where food is a matter of what we can harvest from our backyards?

While simple living is indeed the best solution to our unsustainable living conditions, change does not need to be a drastic measure. Let's look at some of the small ways we can affect real change in our homes and lives.

Sustainable Building Materials

Sustainable building materials can completely transform the initial stage emissions associated with most infrastructural and development projects, greatly impacting the carbon footprint associated with building. These include:

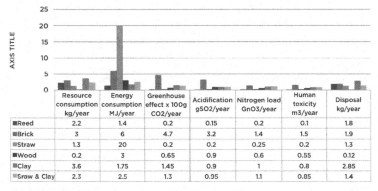

	Resource consumption kg/year	Energy consumption MJ/year	Greenhouse effect x 100g CO2/year	Acidification gSO2/year	Nitrogen load GnO3/year	Human toxicity m3/year	Disposal kg/year
■Reed	2.2	1.4	0.2	0.15	0.2	0.1	1.8
■Brick	3	6	4.7	3.2	1.4	1.5	1.9
■Straw	1.3	20	0.2	0.2	0.25	0.2	1.3
■Wood	0.2	3	0.65	0.9	0.6	0.55	0.12
■Clay	3.6	1.75	1.45	0.9	1	0.8	2.85
■Sraw & Clay	2.3	2.5	1.3	0.95	1.1	0.85	1.4

Figure 7-11: Sustainable building materials can reduce emissions right from the first stage.[42] *Source: Research Gate*

Bamboo

A high-yield and land-use efficient crop, bamboo is a great alternative to wood owing to its high strength-to-weight ratio, which is even greater than concrete and brick. It also takes longer to wear away, making it a great choice for flooring.

Cork

Much like bamboo, cork is a fast-growing, high-yield natural material. Flexible and resistant to high pressure, cork takes a long time to wear away, making it a great alternative to building materials.

Hempcrete

While the usefulness of hemp from an environmental perspective has been well-researched, its use in building and construction is a relatively lesser-known virtue of the miracle plant.

Also known as hemp bricks or hempcrete, this bio-composite building material is a form of vegetal concrete made from the inner, woody core of the hemp plant, which is then mixed with lime and other natural materials to enhance its strength.

Known as the fastest-growing crop on a year-to-year basis, hemp has a naturally **high tensile strength**, making it perfect for construction. Cultivation of the crop also requires no chemical additives, utilizes very little water and contributes to the regeneration of soil quality, making it a boon to both farmers as well as our environment.[43]

Tarun Jami, founder of GreenJams, says:

'Our current homes and offices are making life uninhabitable on Earth. While it's difficult for laypeople to visualize the contribution of our buildings to the changing climate, choosing low-carbon materials such as Agrocrete® and hempcrete also makes economic sense. Since Bio-based construction materials come from a natural biological process, they capture carbon dioxide from the atmosphere, which when combined with the right technology, could remain sequestered forever. This means that the production and usage of these alternative building products helps heal the environment and save money.'

Plant-Based Polyurethane Foam

A type of rigid foam, polyurethane is commonly used in surfboards due to its water-resistant and insulating nature, making it a great alternative to asbestos. Made using natural fibres from bamboo, kelp and hemp, this plant-based building material is quickly picking up in popularity in the western world.

Recycled Plastic

While plastic may well be one of the biggest sources of pollution and environmental degradation known to our times, it can be used for more than just storing fizzy sodas or food.

In a recent trend, manufacturers are diverting plastic waste from landfills by using recycled plastic and other ground-up trash to produce concrete. Recycled plastic can also be blended with virgin plastic to produce polymeric timbres, a great alternative to wood.

Recycled Steel

Unlike most other building materials, steel is 100 per cent recyclable.

The mining process involved in steel manufacturing is highly destructive to the environment, and so making the most optimal use of discarded steel offsets the negative impacts of its production.

While a 2,000-square foot house can take an estimated fifty trees to build, a frame made from recycled steel requires the steel equivalent of only six scrapped cars.[44]

Renewable Forms of Energy

While wind, wave and biofuel energy, among a host of others, can make big changes on an industrial level, there are steps that we, the individual consumers, can take towards cutting down on non-renewable forms of energy. One example of this is solar panelling.

From household electricity to heating and cooling, solar energy can be used for several different needs.

In nations like India, sunlight is ample—the perfect condition for the effective use of solar panelling.

For example, Cochin airport is widely known for being 100 per cent solar-powered. With a solar farm situated just beside the structure, the airport is producing 50,000 to 60,000 units of electricity per day to be consumed for all its operational functions, making the airport power neutral.[45]

Rainwater Harvesting

Rainwater harvesting is a powerful tool to conserve water while also cutting down on utility spending.

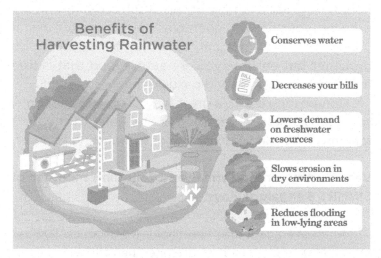

Figure 7-12: Benefits of rainwater harvesting.[46] Source: Treehugger

The practice of collecting and storing rainwater for domestic use, rainwater harvesting provides a source of readily available water in areas where water is scarce, difficult to source or seasonally available. It is also a great source of water for irrigation, livestock and agricultural activities.

Greening Your Spaces

Did you know that research[47] has found that having plants around makes us happier? As we all remember from our fifth grade biology classes, plants are a great source of oxygen, converting CO_2 into breathable air.

Plants also have a cooling effect, meaning that your home will stay comfortable in the summer months.

Greening our homes is a great way to improve our indoor air quality, while also naturally beautifying our homes. A great example of this is terrace farming, i.e., growing your herbs, vegetables and spices for use in your kitchen. This not only reduces your spending on groceries but ensures you are eating food that is free from toxic chemicals and pesticides.

Composting is also a great way to green your home, while simultaneously reducing waste. Refer to chapter 1 for a more detailed explanation of composting at home.

Vani Murthy, urban farmer and founding member of SWMRT Bengaluru, says:

'In an average Indian household, we produce 300 to 400 gms. of solid waste per person per day. Composting presents an unmatched opportunity for every household to combat waste on a person-to-person level. One thing I wish people understood about composting is that it is the most exciting daily action one can take to help mitigate climate change as it is 60 per cent of our waste. It is the magic of mimicking nature. Waste we generate, when composted at the source of generation or at local facilities, would reduce the burden on the municipality and the landfills significantly.'

Green Technology

While technology has indeed exacerbated the decline of our natural world, it has the potential to completely transform the way we live, in harmony with nature. The advent of Internet of Things (IoT) and automated home systems can go a long way toward saving electricity and conserving water.

For example, home temperature control systems can be automated to shut off when the temperature has dropped or risen to its optimal level for comfort, while automated light switches can turn lights on and off based on delicate sensors that can detect whether or not the room is occupied.

Green Planning

An ecologically conscious take on urban and city planning, green planning refers to building vast settlements to decarbonize human activity. This includes planting a large number of trees, the installation of large green spaces and parks, walking and cycling facilities and more.

Living at Home

Whether we know it or not, our home is an ecosystem in and of itself. From the food we eat to how we manage our waste, and from energy to water, heating and cooling, our homes can sustain us completely without us even having to step foot into the world outside.

Making one conscious decision towards revitalizing our living spaces with a focus on the environment can drastically change the way we live overall. After all, we are all a part of the same family, the same home and the same planet.

Resources

Here is an inventory of supplementary media material, people who matter and shops in line with the chapter's theme.

Books

1. *Bare Necessities: How to Live a Zero-Waste Life*
   Author: Sahar Mansoor and Tim de Ridder
   Published by: Penguin Random House
   https://barenecessities.in/collections/books/products/bare-necessities-how-to-live-a-zero-waste-life-book

2. *Sustainable Home: Practical Projects, Tips, and Advice for Maintaining a More Eco-friendly Household*
   Author: Christine Liu
   Published by: White Lion Publishing
   https://www.amazon.in/Sustainable-Home-Practical-maintaining-eco-friendly/dp/071123969X

3. *Imperfectly Zero Waste: A No-Nonsense Guide to Living Sustainably in India*
   Author: Srini Swaminathan Shubhashree
   Published by: Hachette India
   https://www.amazon.in/Perfectly-Zero-Waste-No-Nonsense-Sustainably/dp/9391028446

4. *Made from Scratch: Discovering the Pleasures of a Handmade Life*
   Author: Jenna Woginrich

Published by: Storey Books
https://www.storey.com/books/made-from-scratch

5. *How to Grow Fresh Air: India's Top Experts Teach You How to Beat Air Pollution*
   Author: Kamal Meattle
   Published by: Juggernaut
   https://www.amazon.in/How-Grow-Fresh-Air-Pollution/dp/9386228904

Documentaries

1. *Minimalism: A Documentary About Important Things*
   Directed by: Matt D'Avella
   Available on Amazon
   https://www.amazon.com/Minimalism-Documentary-About-Important-Things/dp/B01KBDQ0FY

2. *The Story of Stuff*
   Directed by: Louis Fox
   Available on Amazon
   https://www.storyofstuff.org

Ted Talks

1. How I Built One of The Most Sustainable Houses in a City?
   By: Snehal Patel
   https://www.youtube.com/watch?v=UuF-4YjXicl

2. A Vision of Sustainable Housing for All Humanity
   By: Vishaan Chakrabarti
   https://www.youtube.com/watch?v=B8kyr
   IQCFXQ

3. Sustainable Living Starts in The Kitchen
   By: Vani Murthi
   https://www.youtube.com/watch?v=hvHEbBnnosI

4. Why I live a Zero-Waste Life
   By: Lauren Singer
   https://www.youtube.com/watch?v=pF72
   px2R3Hg

People

1. Kamana Gautam | @mycocktail_life
   https://www.instagram.com/mycocktail_life

2. Juhi | @mommyingwork
   Sustainable Parenting
   https://www.instagram.com/mommyingwork/
   ?hl=en

3. Sonika | @Sonikabhasin
   https://www.instagram.com/sonikabhasin/

4. Vasuki Iyengar | @Vasuki.Composting
   https://www.instagram.com/vasuki.composting/

5. Soumya | @Greenfeetcleanfeet
   https://www.instagram.com/greenfeetcleanfeet/

Shops

1. Mianzi
   Sustainable Bamboo Products
   Delivers Worldwide
   https://www.mianzi.in

2. Sow and Grow
   Eco-Friendly Gardening Kits
   Delivers PAN India
   https://www.sowandgrow.in

3. Daily Dump
   Compost Bins
   Delivers PAN India
   https://www.dailydump.org

4. Refillable
   Refill Service for Homecare Liquids
   Mumbai| Bangalore| Surat| Pune
   https://refillable.store

5. Beco India
   Eco-friendly Homecare Products Made from Bamboo
   Delivers PAN India
   https://www.letsbeco.com

6. Our Better Planet
   Sustainable Products
   Delivers PAN India
   https://www.ourbetterplanet.com

7. Big Blue Marble
    Sustainable Household Products
    Delivers PAN India
    https://www.instagram.com/bigbluemarble_india

8. Born Good
    Plant-based home cleaning products
    Delivers PAN India
    https://www.borngood.in/

9. Praacheen Vidhaan
    Plant-based products for sustainable living
    https://www.praacheenvidhaan.com/

10. Good Karma
    Recycled tissues, kitchen towels, and other
    products for a sustainable home
    https://www.instagram.com/origamigoodkarma/

8

Beauty and Cosmetics

GHG Impact: 8%[1]

This industry impacts the following Sustainable Development Goals (SDGs):
- SDG 6: Clean Water & Sanitization
- SDG 8: Decent Work and Economic Growth
- SDG 12: Responsible Consumption and Production
- SDG 14: Life Below Water

ACTIVITY

1. How often do you buy makeup products?
   - ☐ Once a month
   - ☐ Once every few months
   - ☐ Other (please specify)

2. How much do you spend on makeup and skincare?
   - ☐ Under Rs 1000
   - ☐ Rs 1000-2000
   - ☐ Rs 2000-3000
   - ☐ Rs 3000-4000
   - ☐ Other (please specify)

3. What feminine hygiene products, if any, do you use?
   - ☐ Pads
   - ☐ Tampons
   - ☐ Panty liners
   - ☐ Menstrual cups

4. Do you follow a fad diet?
   - ☐ Yes (please specify)
   - ☐ No

5. Do you use luxury beauty products?
   - ☐ Yes
   - ☐ No

If there is one thing I have learned after years of interacting with people and relaying my message of sustainability, it is that humans are fragile, predictable species.

Our technological innovations, civil societal structures, supreme intelligence and opposable thumbs would have us believe we were gods with complete dominion over the natural world. But in actuality, we are far from being infallible.

Whether we know it or not, humankind has always existed in a state of perpetual and consensual slavery to one thing—the virtue of beauty.

From the Hindu epic Mahabharata to the famed Trojan War, man has fought battles and razed empires to the ground in the name of beauty, and the all-consuming need to claim it as his own. And it's not just physical beauty, either.

We crave beautiful things like gold and diamonds to wear as expensive baubles around our wrists and necks. We choose to import domestic animals from distant shores as if they were pieces of property and not sentient beings, while a hundred stray dogs bark at our doors with just as much love to offer as their more Instagrammable counterparts. We spend large amounts of federal reserves on the beautification of concrete jungles, having destroyed the natural beauty that once existed there before the trees were replaced with brick and cement.

We do it all in the name of beauty.

From the glamour and glitz of the '50s and '60s to the scantily clad superheroines of the modern age, beauty is the driving force behind silver screen legacies from Hollywood, Bollywood, Tollywood and the rest. Beauty has made history since history was penned by man.

While we have indeed come a long way since the age of the First Men, we are driven by the same instincts and desires. As a part of our genetic makeup, we are evolutionarily drawn to find a suitable mate to reproduce and carry on our genetic lines. To attract this mate, we must demonstrate our superiority over competing suitors.

How do we do this?

The answer is simple: by way of power, or by way of beauty.

But are the two mutually exclusive?

According to research, beauty breeds riskier behaviour, as potential mates compete to seem more powerful, of higher social status or more physically suited to be the best choice. This is a behavioural tactic known as self-signalling.[2]

What this means is that when a heterosexual man is shown a picture of a beautiful woman, he might play a riskier strategy in a board game. When potential mates are watching, people tend to overplay their hand as a show of courage or bravado, like college kids drinking excessively to prove they have the highest tolerance, only to end up leaving the party too quickly due to an upset stomach and a raging hangover.

This is the human behavioural equivalent of rutting behaviours in male deer or mating dances observed in bird species. The idea is to signal that the suitor is both available, as well as strong and capable to protect their object of desire, and any children they may produce over time.

Whether or not we know it, we are hardwired to chase beauty and will go to extreme lengths to claim it for ourselves. While our ancestors may have brawled with other suitors to prove their physical superiority over others, in the modern world, self-signalling does not occur in quite the same way.

If you have ever read the great romances of past ages, you could be familiar with the various material and immaterial analogies for beauty. From a sliver of moonlight to hedonistic pleasures like fine wines, chocolates, flowers, gems and gold, the indescribable value of beauty in our lives is condensed into something tangible. In the modern world, grand gestures like starting a war, or defeating one's competition have been replaced by offerings of these tangible, beautiful things.

I say that we are slaves to beauty because, unlike other aspects of our lives, we have not yet figured out how to control it. Instead, we work all our lives to afford expensive gifts and offerings to appease beauty and invite it into our lives. We barter and trade in sweet words and expensive gifts, hoping that in return, we may finally hold beauty in our hands.

Today, capitalism has found a way to monetize this unspoken, primal desire that is the driving force behind almost everything we do. From expensive treatments that allow individuals to edit almost every physical attribute—whether the lilt of their voice, the colour of their skin, the curve of their smile or the gentle bend of their nose—to the new trend of 15 seconds of fame propagated by social media, today, beauty is bought and sold by the second, and we are as much slaves to it as we have ever been.

In the present era, cosmetics have become a critical feature of our modern lifestyles. The global cosmetics market size was valued at $380.2 billion in 2019 and was projected to reach $463.5 billion by 2027.[3] The global cosmetic surgery market is projected to grow from $46.02 billion in 2021 to $58.78 billion in 2028 at a CAGR of 3.6 per cent.[4]

In the United States, the beauty services sector employs over 6,70,000 people. On a global scale, over 1 million workers are employed in purely the service sector of the industry. To put this into perspective, Statista reports that in 2019, L'Oréal alone employed 87,907 people worldwide.[5]

India's quest for beauty is of great value to its GDP too. In 2020, the market size of the beauty and personal care industry was valued at one trillion rupees. The market size for this industry is projected to increase to an estimated two trillion rupees by 2025.[7]

However, beauty products and treatments paint a pretty picture to hide a dark and storied history of rampant chemical use, animal testing and unsustainability. As the popularity of the products skyrockets, active and inactive residues from the products are introduced into our environment, threatening its natural sanctity.

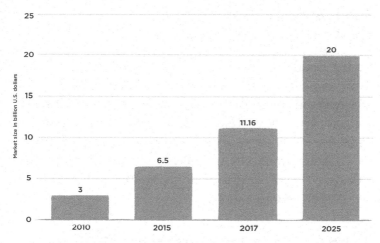

Figure 8-1: Market size of the cosmetic industry across India from 2010 to 2025.[6] Source: Statista

I have never been one to endorse personal beauty. Anyone who has met me will attest that I am a tall, lanky, dusky-skinned individual who is, more often than not, a little dusty from a recent adventure in the great outdoors. To me, true beauty is not a matter of what you see when you look in the mirror, but what pleases your heart. In my eyes, there is nothing more beautiful than inner beauty—one that is defined not by what is on the surface, but rather kindness, empathy and compassion.

As the story of King Midas and his golden touch teaches us, beauty is beautiful, until it is all you are left with. Let's take a closer look at the subject of beauty, our unceasing quest to attain it, and what that means for the natural world.

The Eye of the Beholder:
The Problems with Beauty

Today, beauty is measured not just by the effect it has on others, but rather, by how many choose to fawn over it. Since the dawn of the age of social media, beauty and influence have taken a prominent position in the hearts and minds of children of all ages, prompting an unhealthy cycle of unrealistic standards of beauty, and associated body dysmorphia, that is the warped perception of one's image.

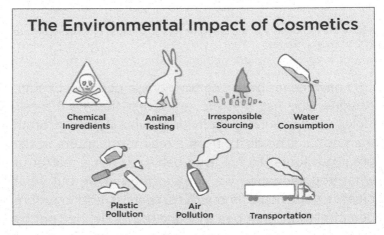

Figure 8-2: The environmental impact of cosmetics.

The cosmetics industry leverages this troubling shift in consumer mindsets to market their products to them and the very same channels that propagate these unrealistic standards, with a recent study by Harvard Business School finding that 62 per cent of women follow beauty influencers on social media.[8] Another study found that 42 per cent of eighteen to twenty-four year olds are inspired by social media when it comes to makeup.[9]

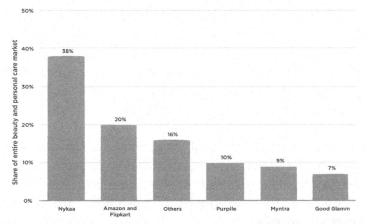

Figure 8-3: Share of online beauty and personal care market in India in 2021, by verticals.[10] Source: Statista

Brands have adapted to the age of user-generated content by investing in influencer marketing, that is paying thought-leaders to promote their products to their audiences through a variety of social platforms. Currently, 86 per cent of the top 200 beauty videos on YouTube are created by users and not brands, while on Instagram, the beauty hashtag dominates with more than 490 million entries.

While this method of marketing is more interpersonal and allows for money to be made by consumers instead of brands, there is a flip side to the practice that often goes unnoticed. As creators and influencers, we are too easily swept away by big brand names and associations and fail to think about what exactly it is we are promoting, and how it could affect not just our followers but the millions of living creatures along the supply chain.

As with most processed food products, most make-up products are made using palm oil—the environmental impact of which I have already discussed in Chapter 1. However, there are a host of other unsustainable ingredients, each with associated impacts on our planet, fellow humans and even animals.

Look at this chart of environmental, social and governance (ESG) risks associated with commodity-based ingredients commonly used in bestselling products in five categories of cosmetics, from face cream to lipstick:

As you can see, each product contains at least one ingredient that is a high-risk commodity. This means that the sourcing, manufacturing, processing or sale of these commodities poses a host of risks to society, people, animals and nature.

This is not unusual for the beauty industry. ESG risks are present across every continent and for every type of cosmetic.

Mined commodities and ingredients are notorious for associated forced labour in producing countries. For example, in our nation, mica, a mineral used to add glimmer and pigment to cosmetics, has been tarnished by its connection with child and forced labour. However, the commodity is still widely used

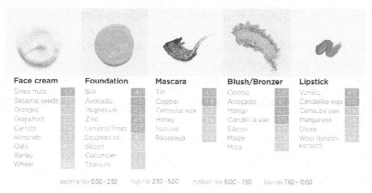

Face cream	Foundation	Mascara	Blush/Bronzer	Lipstick
Shea nuts	Silk	Tin	Cocoa	Vanilla
Sesame seeds	Avocado	Copper	Avocado	Candelilla wax
Oranges	Magnesium	Carneuba wax	Mango	Carnauba wax
Grapefruit	Zinc	Honey	Candelilla wax	Manganese
Carrots	Lemons/limes	Iron ore	Silicon	Olives
Almonds	Soybean oil	Rapeseed	Maize	Wool (lanolin
Oats	Silicon		Mica	extract)
Barley	Cucumber			
Wheat	Titanium			

extreme risk 0.00 - 2.50   high risk 2.50 - 5.00   medium risk 5.00 - 7.50   low risk 7.60 - 10.00

Figure 8-4: Risks associated with commodity-based ingredients in a product.[11] *Source: Verisk*

by cosmetics manufacturers across the globe, being detected in 100 per cent of blush and bronzer products, 60 per cent of foundations and lipsticks and 40 per cent of mascaras.[12]

Issues such as these present themselves along every stage of the supply chain, from ideation to disposal.

Cosmetics supply chains also have a significant environmental footprint, which has brought allegations of deforestation and water pollution.

While it is well known that palm oil cultivation has led to severe deforestation in Indonesia and Malaysia, other natural products and grains such as oats, barley and wheat—often used as extracts in face creams and foundations—have been linked to water pollution in major producing countries, such as France, India, China and Australia.[13]

P(r)etty Plastic—Microbeads, Macro Problems

By now, you are probably familiar with microbeads, and the dire problems they can pose to our environment and health. Plastic microbeads are prevalent across the beauty industry, in body scrubs, face washes, and a host of other exfoliating products.

These microbeads are typically coloured in bright, vibrant shades meant to make the product look more appealing to the end consumer and are marketed as being manufactured using revolutionary new technologies capable of exfoliating the skin on a deep and microscopic level.

In certain recorded cases, plastic has been proven to constitute up to 90 per cent of the ingredients in cosmetic products.[14]

No matter what the expensive marketing campaigns will have you believe in, however, microbeads are just microplastics. They are currently swirling in our oceans, attracting toxins and infiltrating our food chain. Currently, most governmental wastewater treatment plants are not equipped with the filter technologies to remove these microparticles, leading to what some know as the 'Plastic Soup' in our waterbodies.

Although the US and UK have banned the production of microbeads since 2015 and 2018, respectively, their prevalence in beauty products does not reflect this fact.[15] This is because cosmetic manufacturers operating in countries with no microbead regulations have little to no obstacles in their path, and can still export to countries that do.[16]

Not all products made with microplastics are created equal, however, and can be classified as follows based on their environmental impact:

Red	Products in this colour category contain microplastics. Our list of microplastics is derived from the research conducted by UNEP, TAUW, and ECHA. This list contains over 500 synthetic ploymers.
Orange	Products in this colour category contain what we call 'sceptical' microplastics. By 'sceptical microplastic' ingredients we mean synthetic polymers without sufficient information concerning their risks. These include, but are not limited to, Polyquaternium, Polysorbate, PEGs, and PPGs.
Green	Products in this category do not contain any known microplastics or 'sceptical microplastic' from the red and orange categories.

Table 8-1: Microplastics classified based on their impact.[17] *Source: Beat the Micro-Bead*

Unsustainable Resource Consumption

A common misconception among cosmetics consumers is that products that use natural ingredients are sustainable. However, beauty products with natural ingredients are not necessarily better than their chemical-based counterparts.

Palm oil is one of the most used beauty product ingredients, and is a cause for concern because of its effects on the human body as well as the planet. The high demand for palm oil has resulted in widespread and unsustainable cultivation, with palm plantations encroaching upon natural forests and destroying the natural habitats of countless animal species.[18]

Other sources of oils typically used in cosmetics, such as coconut, soy and more, are just as unsustainable as palm when not cultivated in a sustainable manner.

The rampant and gratuitous consumption of resources is not limited exclusively to natural resources either, with non-renewable fossil fuels, metals, and more also adding to the problem.

Many beauty products like wet wipes, sheet masks, cotton pads and more are marketed to be single-use and are discarded immediately after their job is done. We produce 300 million tonnes of plastic each year worldwide, half of which is for single-use items. That's nearly equivalent to the weight of the entire human population.[19]

Chaitsi Ahuja, founder and CEO of Brown Living India, says:

> 'Vote with your wallet: Consumers must ask questions about materials, and processes involved in the creation and testing of the product before it hits the shelves. What we put inside our bodies and on our bodies is a reflection of our value systems and the future we choose. What flows into our water streams and soil is a result of what we put in our bodies and vice versa. A simple thumb rule about clean skincare is this - if you can't put the product in your mouth, you must evaluate if it is actually good to put on your body.'

Wrapped Up With A Bow:
The Hidden Cost of Attractive Packaging

As the saying goes, one should never judge a book by its cover. In the beauty industry, however, the packaging is half the battle won.

Reflecting virtues of status, beauty and luxury, packaging in beauty products, unlike other industries, is more ornate and extravagant than practical, with beauty brands allocating massive budgets towards research into customer perception and glamorous branding.

In most cases, the more expensive a product is, the more elaborate its packaging is. It comes as no surprise then, that about 70 per cent of the beauty industry's total waste comes from packaging, that is 120 billion units every year with paper, plastics, glass and metal waste piling up in landfills all across the world.[20]

Adding to the problem, cosmetic and beauty products, due to their elaborate structuring and branding, are made with a mix of different materials such as plastic, thin metal sheets, paper, glues and more. This makes them difficult to segregate for recycling and presents a high opportunity cost for recycling workers who would rather spend their time recycling waste that is properly segregated.

The Ol' Nip-and-Tuck: Plastic Surgery and Beauty Treatments

Where once plastic surgery was considered an extreme measure only undertaken by the rich and famous, today plastic surgery and non-surgical procedures such as Botox injections are becoming commonplace.

In 2020, the global cosmetic surgery market size was estimated to be $44.55 billion, with a growth projection of $46.02 billion in 2021 to $58.78 billion in 2028 at a CAGR of 3.6 per cent in the 2021–2028 period.[21]

According to the annual International Survey of Aesthetic Plastic Surgery[22], 'Plastic surgery procedures for aesthetic purposes decreased by 10.9 per cent overall in 2020, with 77.8 per cent of surgeons globally experiencing temporary practice closures during the COVID-19 pandemic'. On the other hand, 'Non-surgical procedures (primarily fillers and hair removal treatments) continued to increase, but by lower proportions than seen in previous years (5.7 per cent in 2020, compared to 7.6 per cent in 2019). This resulted in an overall decrease of 1.8 per cent for all procedures.'

According to the survey, the most popular surgical procedures included breast augmentation, which accounts for 16 per cent of all procedures, liposuction at 15.1 per cent, eyelid surgery at 12.1 per cent, rhinoplasty at 8.4 per cent and abdominoplasty at 7.6 per cent. Alternatively, the top five non-surgical procedures included botulinum toxin (known by its trade name Botox). It accounts for 43.2 per cent of all nonsurgical procedures. It is followed by hyaluronic acid at 28.1 per cent, hair removal at 12.8 per cent, nonsurgical fat reduction at 3.9 per cent and photo rejuvenation at 3.6 per cent.

While in 2020, approximately 85 per cent of nonsurgical procedures were performed on women, the procedures most popular among men included eyelid and ear surgeries, liposuction, gynecomastia and rhinoplasty.

From a geographic perspective, the top ten countries for individuals to avail of procedures in 2020 were the USA, Brazil, Germany, Japan, Turkey, Mexico, Argentina, Italy, Russia and India, followed by Spain, Greece, Colombia and Thailand.[23]

However, these procedures are quick-fix solutions to surface-level beauty and do not tackle the underlying issues that may be presenting themselves through our bodily appearance. While most procedures are certified to be safe, there is no lack of evidence of botched surgeries and procedures that doom individuals into looking like comedic caricatures of themselves.

From lip fillers to laser hair removal, beauty treatments and procedures are being marketed for the youth with increasing frequency as commonplace activities. However, this narrative is highly irresponsible and can have life-changing consequences for those who do not properly evaluate the risks associated with such invasive treatments.

But why are so many swayed by these life-changing trends?

Beautiful Inside: The Mental Health Crisis of Beauty

The age of social media has greatly exacerbated body image issues among adolescents, with influencer culture and Instagram likes leading to warped perceptions of their physical appearance. This warped perception of what is beautiful can lead to feelings of shame, depression, anxiety and more.

This trend disproportionately affects young girls, who are typically already struggling with the body and self-image issues associated with puberty.

Let's take a look at some of the issues that arise from this upset balance between self-perception and societal standards, and how they affect us in our most vulnerable state.

Lowered Self-Esteem

According to a survey[24] undertaken of sixty girls, most responded that the images that ruled social media and television were doctored and not representative of the models' or actresses' real bodies. Caucasian girls noted that they felt the opposite gender would evaluate their beauty based on these images, while girls from minority groups indicated that the images 'did not meet the expectations of the reference group they oriented themselves with'.

These societal perceptions create an unachievable standard of standardised beauty, that not only changes the way we view ourselves, but also affects how we believe we are perceived in the eyes of others. To achieve these standards, young girls will often develop extreme responses, such as eating disorders. In 2022, the instances of global eating disorder increased from 3.4 per cent to 7.8 per cent.[25]

Behavioural Changes

While once staring in the mirror and admiring one's beauty was considered a sign of unchecked vanity, today the selfie rules social media platforms and its popularity is increasing every year. A study that compared selfie-takers and non-selfie-takers to evaluate their self-image indicated that those who took more selfies perceived themselves as more attractive and likeable in pictures taken by themselves, rather than those taken by others. This is another example of body image distortion but reflects a more positive self-image.

Unrealistic Beauty Standards

As the saying goes, what you see is not always what you get. In the age of technology, doctored images are commonplace on social media sites, and are virtually indistinguishable from their natural counterparts. From yellowed teeth to belly fat, there is very little that cannot be erased with the right tools.

However, adolescents who are still new to the social media monster do not always have the wherewithal to tell what is real from what isn't. To conform to a standard of beauty that simply does not exist in the real world, they will resort to extreme lengths like crash diets and rigorous or excessive exercise. They are left feeling inadequate when the results do not match what they see online.

Writer Vasudha Rai says:

'The biggest red flag for me is when a brand says they're 100% sustainable. There is nothing like 100% sustainability in consumption. Consumers spend money on beauty products to look better, therefore the selling point has to be the efficacy of the cosmetic. Sustainability must be more than just a sales tactic, or a USP in today's age—it is a matter of essential hygiene. Beauty must come from nature, and be in harmony with nature. There is simply no way to invest in the idea of beauty without investing in our planet.'

Beauty at a Cost: Environmental Degradation

In today's fast-paced industry, beauty trends break the internet and disappear just as quickly, leaving us scrambling to catch up. With every trend, customers shell out money by the millions, hoping to achieve yet another unrealistic standard of beauty before it is replaced by the next one.

Here's how much an average person spends on beauty across the globe every year.

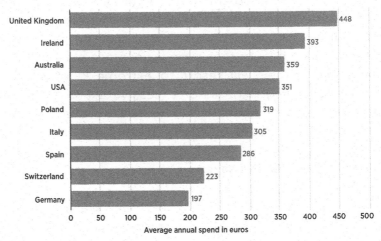

Figure 8-5: Average amount (in Euros) a person spends on beauty across the world. [26] *Source: Statista*

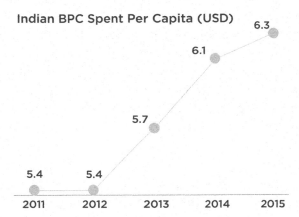

Indian BPC Spent Per Capita (USD)

6.3

6.1

5.7

5.4   5.4

2011   2012   2013   2014   2015

Figure 8-6: Amount spent by Indians on beauty products (in dollars).[27]
Source: Government of India

Comparatively, here's how India spent its money on beauty products over the years:

However, there are also hidden costs to our obsession with beauty, which we, as a global community, cannot afford to pay for long.

Even in some cases of conscious beauty brands, which hero natural ingredients, the carbon footprint associated with the extraction and processing of these ingredients, and packaging and delivery of the products is a cause for concern.

Let's take a closer look at the environmental impacts associated with the beauty industry.

Ocean Chemical Pollution

Whether you are spending a day at the beach or taking a shower in the privacy of your home, toxic chemicals from beauty products leach into our waterbodies via human activity every day. From sunscreen to shampoos, moisturizers and more, beauty products can contain a plethora of chemicals and preservatives like benzophenone, parabens and triclosan, which can be harmful to life when released into the natural world.

These chemicals have been linked with several health issues and have even been linked to cases of cancer.

Oxybenzone, a derivative of benzophenone present in sunscreens, has also been linked with the destruction of coral reefs. According to studies, every year, 14,000 tonnes of sunscreen enter our oceans, and settle on coral reefs, effectively choking and destroying them.[28]

Other harmful chemicals, such as butylated hydroxytoluene (BHT), sodium laureth sulfate, and beta hydroxy acids (BHA), are common in beauty products that leach into our water and threaten to change the biochemistry of marine populations, affecting not only fish but even microscopic plankton on which many species sustain themselves.

Air Pollution

Volatile organic compounds (VOCs) are harmful air pollutants typically found in pesticides, cleaning agents, inks and more. However, they are also prevalent in aerosol-based beauty products, such as hair sprays, perfumes and deodorants.

As with most aerosols, VOCs contribute significantly to GHG emissions, with a recent study demonstrating that household and beauty products emissions contribute to half of VOCs emitted in 33 urban cities.[29]

Deforestation

Palm oil is one of the most used ingredients in cosmetics and beauty products. However, as we discussed in Chapter 1, palm is grown in monocultures, and dense natural forests must be cleared to make place for them.

These forests are home to diverse and endangered flora and fauna that cannot be sustained in palm monocultures.

Packaging in the beauty industry also results in deforestation, as trees are cut to produce raw materials for paper, cardboard, packing tissues, and more—for purely aesthetic materials that are thrown away without a second thought by consumers who purchase the products.

According to studies, packaging made from tree-based materials consumes 18 million acres of forest every year. This widespread and rapid industrial deforestation is a major contributor to global warming, which threatens our very existence on this planet today.[30]

The Ugly Side of Beauty: Exploiting the Defenceless

While environmental degradation is indeed a troubling issue, even more troubling is the issue of exploitation within the beauty industry. Cruel, inhumane acts are carried out all over the world every day in the name of beauty and luxury, and most of us are none the wiser.

The worst part? The affected parties rarely have the tools and means necessary to speak up against the injustices doled out to them, so the cycle continues year upon year, with the number of affected growing day by day.[31]

Ingredient	Properties	Source	Main sustainability aspect
Cetyl alcohol	• May feel tacky when applied to the skin (Baki and Alexander, 2015)	Natural (animal)	Animal exploitation
		Natural (vegetable)	Deforestation (derived from palm oil trees)
		Naturally derived	Petrochemical origin
Octyldodecanol	• Light and lubricious feel of use(Iwata and Shimada, 2013) • Intermediate properties in terms of surface tension and spreadability, with no difficulty of spreading and with high glossiness, and oiliness and leaves a residue after application (Parente et al., 2008, 2005) • Medium polarity • Hydrolytically stable emollient (Alander, 2012)	Natural (vegetable)	Possible deforestation
		Synthetic	Possible pollution due to the heavy metal catalysts (Both et al., 2005)

Table 8-2: Main fatty alcohol emollients. Source: International Society of Aesthetic Plastic Surgery

Ingredient	Properties	Source	Main sustainability aspect
Isopropyl myristate (monoester)	• Intermediate scores of gloss, residue, oiliness and surface tensions and high spreadability (Alexander, 2012: Parente et al., 2008, 2005) • Dry skin feel (Alander, 2012)	Naturally derived synthetic (from animal and vegetable sources)	Animal exploitation and possible deforestation according to the natural source used
Olive oil (natural triglyceryde)	• Reasonably stable against oxidation (Alander, 2012) • Low irritation power • Good cutaneous compatibility	Natural (vegetable)	Decrease of oil production in food industry
Caprylic/ capsule lriglyceride (synthetic triglyceride)	• Intermediate viscosity and polarity • High reistance against oxidation • Medium spreding (Alander, 2012)	Naturally derived synthetic (from vegetable sources- coconut or palm oil)	Deforestation (derived from palm oil trees)

Table 8-3: Main ester-based emollients. Source: International Society of Aesthetic Plastic Surgery

Child Labour and Forced Labour

Many cosmetics products make use of raw materials that must be mined and extracted through processes that are complex and harmful to the environment. However, most of the harm is direct, affecting the labourers themselves.

Unfortunately, in many underdeveloped nations, these labourers are children who are made to work illegally to support their families.

For example, mica, a natural silicate mineral dust found in the Indian subcontinent, is commonly used to brighten and add a shimmer to a variety of makeup products. The mica industry has come under a lot of fire for its reported use of child labourers in the mining of raw materials. A 2016 research study estimated the number of children involved in mica mining in Jharkhand and Bihar at 22,000.[32]

Mining is a dangerous activity for people of all ages and can be especially damaging to children. Besides the life-threatening potential of inhaling particles and fumes, the illegal nature of the activity also prevents the government from insuring them or protecting them from hazards.

In another prominent case, the raw material used for cocoa butter-based face and body creams, has been prominently linked to child labour in the West African country of Cote d'Ivoire, the world's leading producer of cocoa.[33]

The examples don't stop there. From vanilla to shea nuts, copper and silk, ingredients used in everything from lotions to foundations, all have been associated with the exploitation of child labour in the last five years in at least one of their major producing countries.[34]

Mines are no places for children, since workplace accidents are highly common there. According to research, child miners have an average fatality rate of 32 per 1,00,000.[35]

On the other hand, Carnauba wax and candelilla wax, which provide the waxy base for mascaras, has been tainted by associations with labour rights violations in Brazil and Mexico. Here, labourers have been reportedly forced to work long hours and with little to no pay.

Animal Testing

Perhaps one of the best-known dirty secrets of the beauty industry, animal testing, has become the subject of global outrage in recent times.

Testing cosmetics on animals is an inhumane act, and worst of all isn't necessarily reflective of how human beings might react to the same products.

As Cruelty-Free International[36] argues, animal experimentation is a void endeavour, since most animals are not affected by the same illnesses or diseases as humans. To test the effectiveness of a product on a skin condition only known to man, scientists must first artificially infect the animal through the introduction of the disease in a controlled environment, which far from mimics the real-life conditions we humans would face.

While there have been decades of research reporting lab rats and other testing animals recovering from incurable human diseases, the same cure does not necessarily work for humans. Ninety per cent of drugs successfully tested on animals do not work on humans.[37]

Animal testing is also a deliberately cruel act—if the same act were to be carried out on a human subject, the researcher in question would be labelled a twisted psychopath and left to the prison system.

Take skin and eye irritation tests for example— to test a product's irritant ability, animals' eyes and shaved skin are exposed to any number of chemicals. These irritant chemicals can cause painful and uncomfortable reactions, and because the animals are not provided with any pain relief, they must simply suffer in restraints until the experimentation is over, at which point they are either put down or left to recover for a short period before being tested upon again.

Another cruel test that bears the weight of finality is the lethal dose test. Researchers inject animals with a lethal dosage of chemicals to identify and categorise fatally toxic chemicals. This, again, is not necessarily reflective of how these chemicals will react when introduced into the human body.[38]

Adding to the inhumane acts, the animals that are resilient enough to survive all the rounds of testing are usually put down by means of asphyxiation, neck-breaking, or decapitation.[39]

On a lighter note, India is one of the few nations that has made it illegal to use animals for the testing of cosmetic products, even banning the import of animal-tested cosmetic products in 2020.[40]

What Are We Missing?

I have always believed that the modern world is too hung up on how we look and has lost its ability to see beyond what exists on the surface.

But is this truly the only way to live?

Of course not.

I, for one, have never paid much heed to beauty products and treatments, but there were times I too wished I was one of the beautiful ones.

When I was younger, I was teased relentlessly for my appearance. Too skinny, too dark, too wimpy, too quiet—the labels are not unique to my story but did not have the same effect that they do with others I know. As a single-parent child from a middle-class family, frivolous purchases in the name of beauty were not something I even considered growing up. Maybe this is why my interest in the beauty of that kind never blossomed the way it has for others of my generation.

To this day, I have never really done anything for my skin, with the only variables in my appearance being my hair and my clothes, both of which have remained largely the same since I was a teenager.

To me, my appearance is a matter of function. I wear the same four pairs of clothes, even when doing so isn't necessarily considered appropriate. While I sometimes attract suspicious glances and unwanted attention, I have never really minded it much. However, I cannot say the same for those around me.

In the not-so-distant past, I visited a city to attend to some work and was invited to a wedding while I was there. With only my backpack to my name, I was ill-equipped to attend a big fat Indian wedding, but I decided to go anyway.

While others around me were dressed in elaborate sherwanis, kurtas, and embellished *joothis*, I stood clad in my regular uniform—a black t-shirt and black pants. My companion at the time was a fellow traveller, a German man (Ian) who was passing through the city on his way to the next destination. Much like myself, my friend was not prepared to attend a wedding in all its pomp and show and was dressed in linen pants and a worn shirt.

I will never forget that evening because it taught me something very interesting about beauty, and how our perceptions can warp our beliefs.

I recall, clear as day, as we all sat in the room with the men from the wedding party, a man approached me and my friend, sitting there in our travel attire.

'Aakash, get ready quickly! We need to go downstairs!' he said urgently.

The look on his face when I told him I was as ready as I ever would be, was one that makes me laugh to this day. A mixture of shock, surprise and acceptance flashed across his features as he nodded dumbfounded and walked away.

While this interaction was indeed memorable, it is not the man's expression that struck me, but rather his omission of my friend, who was dressed similar to me. I realised then that in this man's eyes, it was acceptable for my friend, a white man, and clearly a traveller, to dress as he did, while far more was expected of me.

This minor interaction may not seem like much but is a telling example of our ability to place people into stereotypical boxes. In the man's eyes, my friend was a foreigner, looking to 'experience Indian culture', while I was just an unbathed Indian man.

However, despite my brown skin, I am the same as my friend insofar as I am a traveller. I do not have a home, or a closet, to call my own. Why then would I invest a significant sum of money into a costume I could only wear on the rare occasions I happened to visit the city when there was a wedding?

I spend much of my life living close to the land, and as anyone who lives close to the land will tell you, the dirt often rubs off, settling on all those who stand too close by. In the village, we are not worried about looking dirty, because we are all soiled to some extent.

While in the city, appearance is considered a sign of social status, beauty is not a priority to a traveller. While it does feel strange and uncomfortable to be in the city, I have the luxury of never having to see these city folk again. Others, I know, are not as lucky.

Life in the city exists in a strange state of flux, as far as beauty is concerned. To illustrate my point, let me tell you another story.

On my recent trip to Delhi, a friend proclaimed, 'Look at Aakash's hair! How did you get it to look so smooth and shiny?'

When I informed her I don't do anything to my hair, besides washing it on occasion, this prompted a conversation about the quality of the water in Delhi, and how damaging it can be to our skin, hair and nails.

'You're so lucky,' my friend said.

But what she did not realize, and what most tend to forget, is that beauty and health are inexorably intertwined. The real reason my skin is so clear, or my hair is so silky, is because I live close to nature. The rivers I bathe from are not polluted or treated with chemicals. The food I eat, the water I drink and the air I breathe are largely free from toxins, while the same can rarely be said for anyone living in the city. This is not just true for me, but for all those who live close to nature. These people age slower, and, in my opinion, are more beautiful than most who have physically altered their features by going under the knife.

Beauty and health are not mutually exclusive but are rather the same thing.

Today, we are sold beauty products, fad diets and treatments under the guise of health. From keto to Atkins to paleo to juice cleanses, social media and the internet are rife with so-called 'experts' proclaiming they have discovered the secret to eternal youth, when in fact, all they have discovered is a quick way to dehydrate yourself or upset your stomach.

According to Samrath Bedi, executive director of Forest Essentials:

'With the absence of an industry-standard definition of "clean", marketing systems are agile enough to accommodate changing consumer preferences to rebrand products to ensure continued marketability for their brand. What is "clean" for one brand may not fit the definition of what makes another brand "clean". A lack of legislation around the true definition of clean beauty often fosters obscurity and vagueness. Ayurveda, however, sets much higher standards than "acceptable" levels of Government regulations, because its basic view on health and its concept of "natural" is at once more encompassing and refined. Also having "tested" its theories for over 6000 years, directly on human subjects, Ayurveda may justifiably claim to know the long-term effects of its treatments and products.'

Whatever little beauty I possess, I owe to nature. I take in the right minerals and the right nutrients, and I owe it not to any miracle supplements or diets, but to how I live.

And it's not just me. Before the age of innovation, India as a nation lived as I do, close to the ground. If you asked your parents or grandparents for solutions to your skin, hair, or other cosmetic woes, they will not offer you chemicals in a tube, or topical creams that come with warning labels. Rather, they will share the wisdom of the earth, and nature. They will tell you perhaps to eat onions for your hair, tomatoes for your skin, exercise in the morning while it's cool and shower with cold water.

Today, beauty is nothing more than a passing fad. With influencers promoting lip injections, invasive treatments, fad diets and more, our perception of beauty changes with every Instagram post, and lasts only if the next one appears on our feeds.

Today, no one cares why the Taj Mahal was built, only that it is beautiful. I would argue that it is beautiful because of why it was built, and not the other way around. Today, we would never expect our lovers and families to build grand structures in our honour. We don't expect more for ourselves because we value ourselves based on how others see us.

To me, however, beauty is far more than skin deep. After all, my face might not be something I can change, but my thoughts, actions, and feelings are.

To me, beauty is empathy. The ability to put yourself in another's shoes, to experience life through their eyes, to be kind, and to forgive—this is what true beauty is. When we learn to respect nature for its beauty, we will see that it grants us beauty in return. When empathy and beauty combine, beauty becomes sustainable.

The world believes in beauty as it is seen and thinks of people as hollowed out on the inside. Our eyes decide what is beautiful based on what they like to look at, but in truth, there is more beneath the surface than we can imagine. If you asked me who the most beautiful person in the world was, I would say the Dalai Lama or Shahid Bhagat Singh.

While my face is a thing made by some external force, I have the power to create beauty through my ability to empathise and look beyond the surface.

I invite you to close your eyes and continue your search for beauty.

I promise you will find it.

A Thing of Beauty: Sustainable Solutions

While the beauty industry is rife with unsustainable practices, small changes in our everyday lives can have a big compounding effect on our planet, as well as on the beauty standards we all hold in such high regard.

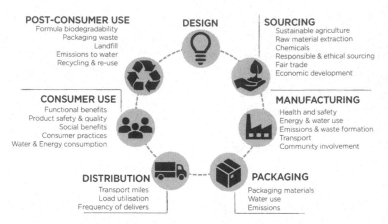

Figure 8-7: Moving towards sustainability in beauty industry.[41] *Source: Journal of Cleaner Production*

Use Fewer Products

The first step to greening your beauty routine is to minimise the number of products you use every day.

Studies have shown that excessive use of beauty products can strip our skin of its natural oils and nutrients, leading to issues like acne, rapid ageing, dark patches and more. While I cannot ask you to abandon all beauty products, minimizing your use of them can make a big difference not just to your appearance, but also to the health of the planet.

Shop From Ethical Beauty Brands

Ethical beauty refers to products and brands that prioritise sustainability and conscious production at every stage of the manufacturing process. From sustainable sourcing of raw materials to ethical labour laws, cruelty-free testing and sustainable packaging, conscious beauty is a force for sustainable change in the world.

Jessica Jayne, founder, Pahadi Local, says:

> 'If you are a consumer looking to combat Greenwashing, here is my advice: Be vocal. Ask more questions. Challenge claims made by brands. Get all the information you need to make a conscious consumer choice. This is the only way greenwashing will make a slow and unwilling surrender.'

In fact, as this graph shows, the demand for conscious beauty has grown by leaps and bounds over the last few years, with more and more people shifting towards beauty products that do not tarnish the natural beauty of our planet.

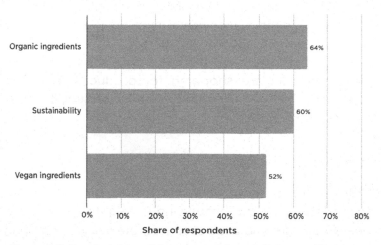

Figure 8-8: Growing demand for conscious beauty.[42] *Source: Statista*

According to Shriti Malhotra, CEO of The Body Shop:

'With the younger consumer demographic driving growth for beauty brands through social media, being a responsible beauty brand is not just the right thing to do, but also a sound business choice. There is now a genuine consumer movement demanding ethical, transparent, eco-friendly and community-positive products that do real good for our planet and its people. The beauty industry has always relied on the "education" route but now, digital-first content marketing means that there is so much more brands can communicate beyond the tutorials, techniques and hacks—it is both an avenue for communication but also a genuine way to spark meaningful conversations, spread awareness, and shine a light on causes and initiatives that brands are supporting. Clean formulations, ethically sourced ingredients, giving back to communities meaningfully and minimizing the environmental impact of the way we do business is becoming the norm.'

Next time you are shopping for products, look for labels such as vegan, cruelty-free or ethical, before making a purchase.

Responsible Disposal

Properly disposing of waste can make a big difference towards making the beauty industry greener. With a heavy reliance on plastic products, the beauty industry is responsible for significant amounts of waste generation.

When throwing out used bottles and containers, be sure to rinse them properly. Recycle wherever possible and limit your use of plastics.

As for the scraps of paper and gratuitous packaging, try and repurpose whatever you can and keep the waste out of landfills for as long as possible.

Change Your Mindset

While these solutions will indeed make a difference in our planet's ecological health, the greatest solution is the one staring right back at us in the mirror. Long before the cosmetics industry existed, people of all genders would still see beauty in one another, and life would go on as it always has.

To change the beauty industry, we must first change our perception of beauty itself. Remember, you don't need to change a thing about yourself to feel or to be beautiful. Look for beauty in the words and actions of yourself and those around you, and not what exists on the surface.

This is where true beauty lies.

Resources

Here is an inventory of supplementary media material, people who matter and shops in line with the chapter's theme.

Books

1. *Glow Indian Foods, Recipes, and Rituals for Beauty, Inside, and Out*
    Author: Vasudha Rai
    Published by: Penguin Random House
    https://penguin.co.in/book/glow

2. *No More Dirty Looks: The Truth About Your Beauty Products*
    Author: Alexandra Spunt and Siobhan O'Connor
    Published by: Da Capo Lifelong Books
    https://www.amazon.in/Dirty-Looks-Siobhan-CDATA-OConnor/dp/0738213969

3. *Sustainable Beauty: Practical Advice and Projects for an Eco-Conscious Beauty Routine*
    Author: Justine Jenkins
    Published by: White Lion Publishing
    https://www.amazon.in/Sustainable-Beauty-Practical-projects-eco-conscious-ebook/dp/B09NGK9V6H

Documentaries

1. *The Truth About Makeup*
    Directed by: BBC

Available on YouTube
https://www.youtube.com/watch?v=aUalQYLmi6U

2. *Toxic Beauty*
    Directed by: Phyllis Ellis
    Available on Amazon
    https://www.amazon.com/Toxic-Beauty-Phyllis-Ellis/dp/B0846VDB96

3. *Broken (Makeup Mayhem)*
    Directed by: Sarah Holm Johansen
    Available on Netflix
    https://www.netflix.com/in/title/81002391

4. *The Dark Secret Behind Your Shiny Makeup*
    Directed by: CNA
    Available on YouTube
    https://www.youtube.com/watch?v=LS_CR7UwhRs

People

1. Marta | BottegaZerowaste
    Zero Waste Beauty
    https://www.instagram.com/bottegazerowaste

2. Preiti | Preitibhamra
    Sustainable and cruelty-free Beauty
    https://www.instagram.com/preitibhamra

3. Anya Gupta | @anya.gupta
    https://www.instagram.com/anya.gupta/

Shops

1. Juicy Chemistry
   https://juicychemistry.com/

2. Pahadi Local
   https://pahadilocal.com/

3. Asa Beauty
   https://asabeauty.com

4. Disguise Cosmetics
   https://www.disguisecosmetics.com

5. Bon Organics
   https://www.bonorganics.com

6. Vaseegrah Veda
   https://vaseegrahveda.com

7. The Tribe Concepts
   https://thetribeconcepts.com

8. Earth Rhythm
   https://earthrhythm.com/

9. Pure Earth
   https://india.purearth.asia/

10. Ruby's Organics
    https://rubysorganics.in/

11. Maati Care
    Plant-based Haircare, Skincare, and Superfoods
    https://maaticare.in/

12. Kastoor
    Plant-based Indian Perfumery
    https://www.kastoor.co/

9

Plastic

GHG Impact: 3.4 per cent[1]

This industry impacts the following Sustainable Development Goals (SDGs):
- SDG 3: Good Health & Well-Being
- SDG 6: Clean Water & Sanitization
- SDG 9: Industrial Innovation and Infrastructure
- SDG 12: Responsible Consumption and Production
- SDG 13: Climate Action
- SDG 14: Life Below Water
- SDG 15: Life on Land

ACTIVITY

1. How often do you consume fizzy drinks?
   - ☐ Once a week
   - ☐ Twice a week
   - ☐ More than twice a week (please specify)

2. How often do you segregate your waste?
   - ☐ Once a week
   - ☐ Twice a week
   - ☐ More than twice a week (please specify)

3. Do you store your plastic bags for reuse?
   - ☐ Yes
   - ☐ No

4. Do you carry a cloth bag when you go grocery shopping?
   - ☐ Yes
   - ☐ No

5. Does your locality have a recycling plant?
   - ☐ Yes
   - ☐ No

As human innovation goes, there is one invention that has transformed the way we live more than any other.

While the wheel has allowed us to travel and see the world, technology and the internet have connected us across borders and barriers of language, commercial agriculture has removed us from the rungs of the food chain and democracy has afforded us the god-like power to judge morality and maintain civility— there is one innovation that still reigns supreme, even surpassing us humans as the self-proclaimed rulers of the modern world: plastic.

Plastic has changed the way we live and will outlive us all. It dictates how industries work, how we eat, what we consume and eventually, if we're not careful, how our world will end.

From plastic bags to packaging, straws, bottles and more that choke our oceans and smother the air, plastic waste has become the subject of global attention in the fight for sustainability. But how did we get here?

Once, plastic was a miracle material, promising a brighter future for all of humanity. An infinitely malleable material capable of storing liquids of all kinds, and withstanding impacts associated with large-scale transportation, plastic offered a cheap alternative solution to some of the goods and services industries' biggest logistical troubles. Whether it's made to be thick, thin, clear, opaque, strong or weak, there is very little that cannot be achieved with plastic. To this day, the material represents infinite possibilities, and is both a blessing and a curse, depending on how it is used.

In 2021, the global plastics market was valued at $593 billion and was projected to grow to reach a value of more than $810 billion by 2030, registering a CAGR of 3.7 per cent.[2]

Cumulative global production of plastics
Plastic production refers to the production of polymer resin and fibers.

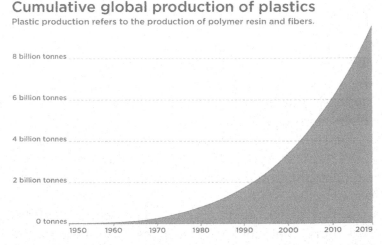

Figure 9-1: Plastic production touched eight billion tonnes in 2019.[3]
Source: Our World in Data

Global plastics production was estimated to be 367 million metric tonnes in 2020 and decreased by approximately 0.3 per cent due to the coronavirus pandemic.[4]

While this minor decrease was indeed a promising trend, it did not even make a dent in the larger issue looming over our planet. Annual global production of plastics has increased more than 200-fold since 1950.[5]

In 2015, the world produced more than 380 million tonnes of plastic. For context, this is roughly equivalent to the mass of two-thirds of the world's population. By 2015, cumulative plastic production was more than 7.8 billion tonnes. This is equivalent to more than one tonne of plastic for every person alive today. Adding to this, plastic on average takes more than 400 years to degrade by natural mechanisms, meaning that most of it still exists in some form in our natural world.[6]

Plastic is also one of the most well-travelled of all pollutants, going where no man has gone before. To illustrate, plastic has been detected at the bottom of the Mariana Trench, the deepest part of the ocean known to man, as well as Henderson Island, a UNESCO World Heritage Site with no human inhabitants. However, over 37 million plastic fragments were found here. In 2019, ten tonnes of trash was recovered from perhaps one of the least accessible places known to man—Mount Everest, the highest point in the world.[7]

Today, plastic is ubiquitous in the natural world. It is so common, that it is becoming a part of our planet's fossil record as a marker of the Anthropocene, with a new name being given to a newly emerged marine microbial habitat called the plastisphere.[8]

With so much plastic floating around our oceans, piling up in landfills and releasing toxic fumes in burning garbage piles across the globe, plastic pollution is perhaps the single biggest threat to human health and our planet's ecology. It results in a plethora of hidden costs for the global economy.

As an advocate for sustainability, I have made plastic the core focus of many of my projects. I have scoured the raging waters of the river Ganga to recover tonnes of plastic waste and have plucked plastic seeds of unsustainability from otherwise pristine mountainsides, valleys and forests across our nation.

Let me tell you the story of plastic, and how what once seemed a miracle delivered by divine intervention, soon transformed into the devil's work.

Problems: Red Flags and Plastic Bags

Plastic pollution and its effects have been well-documented. It has also resulted in a global movement calling for a reduction in plastic production and utilisation. However, in our conversation about plastic, we have forgotten to acknowledge the industry's secret godfather that fuels its growth across the board and will do anything to prevent losing its indirect hold over the world—the fossil fuel industry.

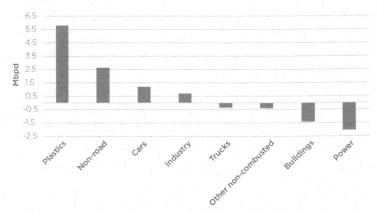

Figure 9-2: Oil demand and growth 2020--2040.[9] Source: Carbon Tracker

That's right. Whether you are consciously aware of it or not, plastic has its roots in the most influential industry of our times and is produced as a by-product of the waste generated from the processing of oil and coal. Oil used for plastics represents two-thirds of total oil demand in the petrochemical sector. Moreover, data from BP implies that this share will rise over time to reach 77 per cent in 2040.[10] This means that the industry most directly contributing to global warming is also the industry profiting from the act of choking our planet.

And the worst part? They will stop at nothing to ensure their pockets remain filled.

Plastic has infiltrated our lives across the planet, and in more ways than we can even imagine. But before we talk about the problem on a global scale, let's take a look at our own backyards, and pick up the littered pieces of the puzzle to tie it all together.

India and Plastic

India's relationship with the plastic industry dates back to 1957 with the dawn of polystyrene production on our shores. Since then, the industry has grown rapidly, spanning the length and breadth of the country.

Today, the industry employs more than 40 lakh people and constitutes 30,000 processing units. Among these, 85–90 per cent belong to small and medium enterprises. India manufactures a variety of plastic products and products that use plastic to some extent. These include everyday plastic goods, medical items and equipment, houseware, fishnets, pipes, raw materials and more.

With more than 2000 exporters around the nation, our major plastic exports include raw materials, woven sacks, fabrics, films, tarpaulin and sheets.[11]

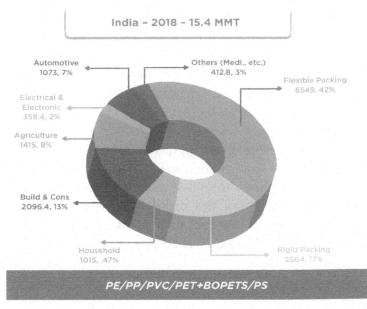

Figure 9-3: Plastic trends in the industry.[12] Source: Centre for Financial Stability/

Plastic production in India has surged over the past 50 years—from 15 million tonnes (MT) in 1964 to 311 MT in 2014. The cumulative exports of plastics and related materials during 2021–22 was valued at $13.34 billion, a 33.4 per cent increase from the 2019–20 exports which were valued at $10 billion. Plastic raw materials were the largest exported category and constituted 30.7 per cent of the total exports in 2021–22, recording a growth of 26.5 per cent over the previous year. Plastic films and sheets were the second largest category, comprising 15.2 per cent of the total exports, and grew by 32.7 per cent over the previous year.

Within the nation, plastic was mostly used for packaging products, with a number standing at 42 per cent, while building and construction was the second largest sector utilising 19 per cent of the total.[13]

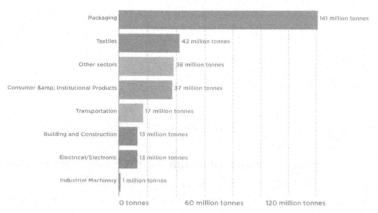

Figure 9-4: Plastic waste generation by industrial sector, 2015.[14] Source: Our World in Data

India is also home to several Plastic Parks, that is, industrial zones developed in clusters to consolidate and synergize the capacities of the domestic downstream plastic processing industry to boost employment and attain environmentally sustainable growth. Currently, ten Plastic Parks have been approved by the Department of Chemicals and Petrochemicals. Among these, six have received final approval from Madhya Pradesh (two parks), Assam

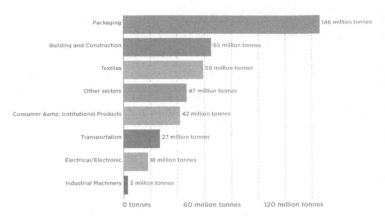

Figure 9-5: Primary plastic production in industrial sector, 2015.[15] Source: Our World in Data

(one park), Tamil Nadu (one park), Odisha (one park) and Jharkhand (one park).[16]

The Government of India is investing greatly in the plastic sector, with plans to take the industry from its current Rs 3 lakh crore ($37.8 billion) of economic activity to Rs. 10 lakh crore ($126 billion) in four to five years.[17]

While it is clear that the plastic industry is a profitable for India, the growing global and local demand is greater than the resources we have to effectively dispose of plastic waste or recycle it. In the financial year 2019, the total demand for major plastics across India was approximately 16 million metric tonnes. According to the Central Pollution Control Board report 2019–2020, India recycles around 60 per cent of plastic waste. The remaining 40 per cent ends up in landfills, on the streets, clogs water bodies, etc.[18]

Another report states that India consumes an estimated 16.5 million tonnes of plastic annually; this

Plastic problem

India's four metros generate more than 1,670 tonnes of plastic waste per day, which is over 40 per cent of the plastic waste produced in India's 60 major cities*

Srinagar
28.14 (5.12%)

Jammu ●
21.68 (7.23%)   Shimla
Amritsar ●   2.23 (4.45%)
24.42 (4.44%)
Chandigarh ●
8.18 (3.1%)
Delhi Dehradun
408.27 (10.14%) 14.56 (6.67%)
Faridabad   Agra
79.03 (11.29%) 40.89 (7.86%) Lucknow   Guwahati
●   70.84 (5.9%) Patna   10.27 (5.04%)
Jaipur   Kanpur   12.6 (5.73%)   Shillong
15.58 (5.03%) 106.66 (6.67%)   5.27 (5.44%)
Varanasi   Dhanbad
Bhopal   25.92 (5.76%) 7.52 (5.02%)   Agartala
23.08 (6.59%)   5.83 (5.71%)
Ahmedabad   Ranchi
241.5 (10.5%)   Raipur 8.29 (5.92%) Kolkata
Surat   23.76 (10.61%)   425.72 (11.6%)
149.62 (12.47%)
Mumbai ●   Bhubaneswar
408.27 (6.28%) ●   51.92 (7.98%)
Pune
101.35 (7.8%)   Hyderabad
199.33 (4.75%)

Vijaywada
43.72 (7.29%)

Chennai
429.39 (9.54%)
Bengaluru
313.87 (8.48%) Puducherry
26.46 (10.46%)
Kavaratti Coimbatore
0.24 (12.09%) 66.31 (9.47%)
● Thiruvananthapuram
15.06 (6.02%)

Legend
● Name of city
XX Plastic waste (tonnes per day)
(XX%) Plastic waste (% of municipal solid waste)
* Of the 60 cities covered in the source, this map depicts 35. The cities have been chosen to depict a cross-country picture

Average plastic waste generation in India (tonnes per day)	Average plastic waste share in municipal solid waste in India
4,059.18	6.92%

Figure 9-6: Plastic generation across India.[19] Source: Down to Earth

would full 1.6 million trucks. Of this, 43 per cent is single-use plastic that ends up in garbage bins. In all, 80 per cent of the total plastic produced in India is discarded.[20]

While the Indian government's ban on single-use plastics is indeed a step in the right direction, there are many loopholes to what may be considered single-use, which keeps much of the problematic plastic in play around the nation.[21]

STOP PLASTIC
WHAT'S TO BE PHASED OUT

	NAME OF THE PRODUCT	UTILITY INDEX	ENVIRONMENTAL IMPACT
	Carry bags thin (less than 50 microns)	32	84
	Non-wovwn carry bags and covers (less than 80 gsm and 320 microns)	21	87
	Straws / Stirrers	16	87
	Small Wrapping / Packaging films	22	84
	Cutlery: Foamed cups, bowl and plates	23	91
	Cutlery: Laminated bowls and plates (non-foamed)	25	88
	Cutlery: Small plastic cups/containers (less than 150 ml and 5g)	23	85
	Ear buds and plastic sticks for balloons, flags, candies etc	17	89
	EPS for decoration	23	85
	Plastic banners (less than 100 microns)	22	64
	Cigarette overlap film	46	62
	Disposable rigid cups, trays and containers	46	56
	Wrapping films for food applications*	64	56

*Only sweet boxes with wrapping films are listed for phasing out

STOP PLASTIC
WHAT'S NOT TO BE PHASED OUT

NAME OF THE PRODUCT	UTILITY INDEX	ENVIRONMENTAL IMPACT
Cigarette filters (non-biodegradable)	20	93
Small plastic bottle for drinking water (<200 ml)	26	79
Plastic bottles for non-food applications	59	40
Plastic bottles for food and beverage (more than 200 ml)	74	36
Multi-layered packaging (more than 36 cm2)	81	73

Figure 9-7: The kind of plastic that needs to be phased out.[22] Source: Down to Earth

A living example of India's mismanagement of plastic waste looms taller than the Taj Mahal, a 73-metre-tall (240 feet) marble structure and is situated in Agra about 240 kilometres east of our nation's capital. The famed Ghazipur trash dump is just one of many examples of giant mountains of trash that litter our great nation.

In the financial year 2019, Madhya Pradesh had 378 landfills. It was the highest number in any state that year, with Maharashtra, Karnataka, and Rajasthan following close behind.[23]

While a lack of infrastructure is indeed to blame, the fault is not entirely the government's or the consumer's, but the products themselves. This is because not all plastic is made equal, making the matter of recycling a dilemma for even the most developed countries.

But what exactly are these factors that differentiate one type of plastic from the other?

One Size Doesn't Fit All: Types of Plastic

While recycling is often presented as our last line of defence against the age of unsustainability, in truth, not all plastic can be completely recycled.

There are mainly seven types of plastic, each with an associated numerical code for easy identification. Of these, only those coded 1 and 2 are 100 per cent recyclable plastics.

Let's look at these in further detail, to understand the complexities of recycling plastic products.

1: Polyethylene Terephthalate (PET or PETE)

PETE plastics are some of the most commonly encountered plastics in our daily lives. Characterised by their lightweight, resilient, and generally transparent nature, PETE plastics are used to package food and the manufacture polyester-based fabrics.

Sources: Beverage bottles, food packaging, jars, polyester clothing, polyester-based rope.
Recyclability: Easily recyclable up to 10 times.[24]

2: High-Density Polyethylene (HDPE)

In technical terms, polyethylene is the most prevalent plastic found across the world. However, it can be

Type	Category	Examples	Recyclable
Thermoplastics	PS (Polystyrene)	Foam hot drink cups, platic cutlery, containers and yogurt	Partially
	PP (Polypropylene)	Lunch boxes, take-out food, containers, ice cream containers	Partially
	LDPE (Low-density polyethylene)	Garbage bins and bags	Partially
	PVC (Plasticized polyvinyl chloride or polyvinyl chloride)	Juice or squeeze botels	Yes
	HDPE (High-density polyethylene)	Shampoo containers or milk bottles	Yes
	PET (Polyethylene terephthalate)	Fruit juice and soft drinks bottles	Yes
Thermoset and others	Multi-layer and laminated plastics, polyurethane foam, Bakelite, polycarbonate, melamine, nylon etc	Car parts, mattresses, circuit boards and electrical insulators	No

Figure 9-8: What kind of plastic gets recycled?[25] *Source: Centre for Science and environment*

classified under three distinct categories: high-density, low-density and linear low-density.

Code 2 plastic refers to high-density polyethylene, a tough, moisture- and chemical-resistant type of plastic that is ideal for the making of cartons, containers, building materials and more.

Sources: Tetra packs and cartons, detergent bottles, toys, buckets, plastic benches, strong piping, etc.
Recyclability: Recyclable up to 10 times.[26]

3: Polyvinyl Chloride (PVC or Vinyl)

A versatile plastic, PVC can be rendered hard and rigid, or soft and flexible, making it ideal for diverse uses. It is also a bad conductor of electricity, can be disinfected quite easily and is impervious to germs, giving it many applications in healthcare.

However, PVC is also considered the most dangerous plastic known to humanity. It is known to leach dangerous toxins such as lead, dioxins and vinyl chloride throughout its life cycle.

Sources: Plumbing and electrical pipes, credit cards, toys, rain gutters, plastic furniture, intravenous fluid bags, medical tubes and oxygen masks.
Recyclability: Sometimes recyclable

4: Low-Density Polyethylene (LDPE)

LDPE is the softer, more flexible and transparent cousin of HDPE. This variety of plastic is a versatile polyethylene known for its ability to lend a thin layer of protection to corrosion-resistant work surfaces and other products.

Due to its cheap and low-quality nature, LDPE is typically not financially viable to be recycled, and will typically be rejected by municipal recycling centres.[27] This is also because, thin LDPE plastic bags have been known to get entangled in the recycling machinery, threatening the entire operation. However, it can technically still be recycled into bin liners and other thin plastics.

Sources: Cling wrap, zip lock bags, bubble wrap, beverage cups, and garbage and grocery bags.
Recyclability: Sometimes recyclable

5: Polypropylene (PP)

Known as one of the most durable types of plastic available in the market, PP is more heat resistant than its counterparts and is flexible enough to endure some slight pressure while still maintaining its overall shape. This combination of virtues makes PP ideal for the packing of hot meals.

However, the practice of recycling PP is complex and costly. Typically used for food deliveries, PP tends to hold on to the smell of the items stored within it, and usually ends up black or grey post recycling, making it difficult to reuse in branded packaging. For these reasons and more, recycled PP is used for industrial applications such as the manufacturing of auto parts, lumbers and more.

Sources: Straws, CD cases, hot food containers, tape, bottle caps, disposable diapers, etc.
Recyclability: Sometimes recyclable

6: Polystyrene (PS or Styrofoam)

Styrofoam is a low-cost, well-insulated plastic that is a favourite in the food and packaging industry. However, PS, like PVC, is a toxic plastic, known to leach toxic chemicals like styrene, a known neurotoxin, which can easily then be absorbed by the food items and ingested.

Polystyrene typically can't be recycled at municipal recycling plants, but rather must be transported to a centralized and dedicated plant. This process is typically expensive, meaning PS usually goes straight to the bin. Recycled PS is also unusable for products that contain or come into contact with food due to health concerns.

Sources: Disposable cups, take-out containers, shipping packaging, foam peanuts, egg cartons, cutlery and building insulation.
Recyclability: Sometimes recyclable

7: Other

The final category of plastics accounts for every other type of plastic that doesn't fall under the first six categories or are combinations of them. Due to the mixture of different types of plastics that exist under code 7, these plastics are typically not recycled.

Sources: Eyeglasses, baby and sports bottles, electronics, CD/DVDs, lighting fixtures and clear plastic cutlery.
Recyclability: Rarely recycled

Single-use plastic products exist across these categories, meaning that each of these categories produces a product that is used only once before being discarded, or, in some lucky cases, recycled.

To recycle any type of plastic, we must first segregate and gather them by their codes —these numbers are typically printed on the containers. While types 1 and 2 are usually the simplest to isolate, others can be more challenging, containing residue from the items contained within them.

When the process of segregation is mismanaged or isn't undertaken with care, that is plastic bottles of fizzy drinks or food packaging being thrown away without rinsing, this can cause major problems for

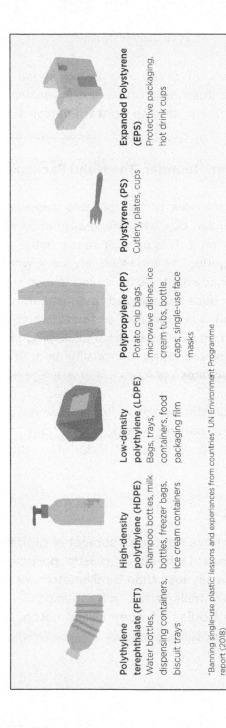

Polyethylene terephthalate (PET)
Water bottles, dispensing containers, biscuit trays

High-density polythylene (HDPE)
Shampoo bottles, milk bottles, freezer bags, ice cream containers

Low-density polythylene (LDPE)
Bags, trays, containers, food packaging film

Polypropylene (PP)
Potato chip bags microwave dishes, ice cream tubs, bottle caps, single-use face masks

Polystyrene (PS)
Cutlery, plates, cups

Expanded Polystyrene (EPS)
Protective packaging, hot drink cups

"Banning single-use plastic: lessons and experiences from countries" UN Environment Programme report (2018)

Figure 9-9: Types of plastic and recyclability factor.[28] Source: United Nations Development Programme

the machinery involved. To avoid these troubles, recycling facilities will often dump entire bags of segregated plastic due to contamination from other substances.

Prianka Jhaveri, founder, The Mend Packaging says:

> 'Most brands think of "sustainable packaging" as a flimsy brown box with a non-aesthetic, minimal black print on it. This perception is a battle we are actively fighting. At the Mend, we work with local waste management partners and paper mills to locally produce customized products in vibrant colours and various shapes and sizes, which are as sturdy and affordable as virgin packaging options, without compromising on quality and at more competitive rates.'

While solid plastic waste is known for its damaging effects on our natural world, plastic can also affect us on another level, one that is far less visible to the naked eye.

Microplastics

Microplastics have become a subject of global concern in recent times. Defined as plastic particles with a diameter typically less than 5 millimetres, or on scales less than 4.75 millimetres, microplastics have been found in the bodies of commercial catch, bees and even unborn foetuses.

Figure 9-10: Hotspots of plastic waste mismanagement.[29] Source: Our World in Data

Figure 9-11: Annual emission from plastic lifecycle.[30] Source: Plastic Pollution Coalition

Particle category	Diameter range (mm = millimetres)
Nanoplastics	< 0.0001 mm (0.1Qm)
Small microplastics	0.0001 – 1 mm
Large microplastics	1 – 4.75 mm
Mesoplastics	4.76 – 200 mm
Macroplastics	>200 mm

Figure 9-12: Microplastics and their size[31]. Source: Our World in Data

Smaller particles, measuring less than 0.0001 millimetres in diameter are referred to as nanoplastics.

Microplastics come from a variety of sources and can be attributed to several primary and secondary processes. Primary microplastics are manufactured to be minuscule, such as microbeads in cushions, toothpaste, detergents, face washes, pellets and capsules. Secondary microplastics result from the wearing away of larger plastic products.

When macroplastics are exposed to friction, either manufactured or in the natural world in rivers and oceans, weathering occurs, breaking the plastics down into smaller and smaller particles.

These microplastics are difficult to track and monitor and can enter our waterbodies, meat and seafood. They can even be inhaled, thus entering our bodies as well as those of sensitive wildlife.

However, the majority of macroplastics (82 million tonnes) and microplastics (40 million tonnes) resurface, wash up, and are buried along the shorelines of the world. According to surveys, most macroplastics have been present on our shorelines for the last 15 years with some dating even further back into our history. This troubling revelation is indicative of the slow degradation of plastics in our natural environment.[32]

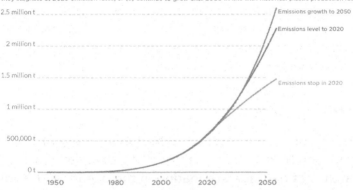

Microplastics in the surface ocean, 1950 to 2050

Microplastics are buoyant plastic materials smaller than 0.5 centimeters in diameter. Future global accumulation in the surface ocean is shown under three plastic emissions scenarios: (1) emissions to the oceans stop in 2020; (2) they stagnate at 2020 emission rates; or (3) continue to grow until 2050 in line with historical plastic production rates.

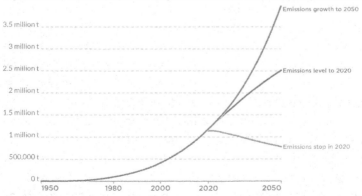

Macroplastics in the surface ocean, 1950 to 2050

Macroplastics are buoyant plastic materials greater 0.5 centimeters in diameter. Future global accumulation in the surface ocean is shown under three plastic emissions scenarios: (1) emissions to the oceans stop in 2020; (2) they stagnate at 2020 emission rates; or (3) continue to grow until 2050 in line with historical plastic production rates.

Figure 9-13: Macroplastics in the surface ocean, 1950 to 2050.[33] Source: Our World in Data

In offshore environments, macroplastics can be traced back even further, having accumulated there for longer. According to the survey, some macroplastics can be traced as far back as the 1950s and 1960s. Most microplastics, approximately three-quarters of all those found in offshore environments, are from the 1990s or earlier.

These large and small, old and new plastic particles have one thing in common—no matter what type of plastic they may be, they are harmful to our environment in more ways than one.

Plastic Pollution

Plastic pollution affects our planet in a variety of ways, from solid waste collecting in landfills to floating patches of garbage stagnating in the ocean, and from micro and nano plastic particles infiltrating our bodies to the toxins building up in our air from unregulated burning.

Every day, we produce plastic on a massive scale sometimes without even realising it. To illustrate, cigarette butts with minuscule plastic fibres are the most common plastic waste detected in the natural world, followed by food packaging, plastic bottles, plastic bags, straws, stirrers and more.

Of the over seven billion tonnes of plastic waste generated globally to this date, less than 10 per cent has been recycled. Every day, millions of tonnes of plastic waste are abandoned in the natural world or shipped to underdeveloped nations to be burned or buried. Besides the waste being generated and the

Macroplastics are greater than 0.5cm in diameter
Microplastics are smaller than 0.5cm

Shoreline Total from 1950 to 2015:
Dry lands   82M tonnes macroplastic
            40M tonnes microplastic

Two-thirds of buoyant macroplastic released into the marine environment since 1950 is stored close to the oceans' shorelines.

A large part of the 'missing plastic' problem is explained by plastic accumulation, burial and resurfacing along shorelines.

Accumulated plastic by age (tonnes)

Coastal Total from 1950 to 2015:
Shallow waters (<200m)   150,000 tonnes macroplastic
                          80,000 tonnes microplastic

Most macroplastic (79%) in the coastal environment is less than 5 years old.

Accumulated plastic by age (tonnes)

Offshore Total from 1950 to 2015:
Deep waters (>200m)   1M tonnes macroplastic
                       0.5M tonnes microplastic

In offshore environments, older macroplastics have had longer to accumulate- plastics younger than 5 years accounts for only 26%

Macroplastics older than 15 years old account for nearly half.

Most macroplastic (74%) is from the 1990s and earlier, suggesting longer breakdown timescales.

Accumulated plastic by age (tonnes)

Figure 9-14: *Where does plastic accumulate in the ocean?[34] Source: Our World in Data*

harm associated with the same, the approximate loss associated with sorting and processing is $80–120 billion.[35]

Feminine Hygiene Products

Feminine hygiene products are a major source of environmental destruction and deserve their own section in this chapter.

Feminine hygiene products such as sanitary napkins, tampons and panty liners are necessary for all adolescent menstruating people across the world. However, only a year's worth of a typical feminine hygiene product used by just one person carries a carbon footprint of 5.3 kilogram CO_2 equivalents.

In developing nations, as knowledge of feminine hygiene increases, sanitary napkins are being used more and more. A study conducted in the slums of Hyderabad reports an increased use of sanitary pads (56 per cent), suggesting that development initiatives have percolated down to the urban poor.

While sustainable alternatives exist, their popularity has been growing at a sluggish pace, and without proper disposal mechanisms in place, this problem promises to compound significantly over the coming years.[36]

Environmental Impacts

The plastic waste that enters rivers and lakes finds its way into the open ocean. According to the UN, it is estimated that 1000 rivers are accountable for nearly

80 per cent of global annual riverine plastic emissions into the ocean, ranging between 0.8 and 2.7 million tonnes per year, with small urban rivers being the most polluted.

Microplastic waste can also enter our soil, being leached through water or becoming trapped in dirt. Degrading plastics in the soil can absorb toxic chemicals like polychlorinated biphenyl (PCBs) and pesticides like dichlorodiphenyl trichloroethane (DDT) eventually making their way into the food chain.

The breakdown of plastics in soil and water also releases toxic chemicals like phthalates and Bisphenol A (BPA). These cause cancer and affect fertility and hormone release in humans. Plastic waste may also clog sewage systems, leading to the spread of diseases and allowing mosquitos and other pests to breed.[37]

Pradeep Sangwan, founder, Healing Himalayas, says:

'Enough data is available that suggests the reasons why Himalayan glaciers are the fastest retreating glaciers in the world—Open-air waste burning releases major pollutants, freshwater pollution increases dependency on PET water bottles leading to further pollution, and over-consumption, and topographical configuration only adds to the worries. In a nation of 50 million people, not a single community, village, or district can say that they have an effective, sustainable solid waste management system. Concerned departments and Panchayats must prepare mentally for an endless war against solid waste due to the disconnect between people and the natural world.'

Indigenous Communities

Plastic also impacts indigenous communities disproportionately. There have been several instances in which indigenous peoples and local communities have been expelled from their lands to make way for mining and industrial purposes.

These communities are highly dependent on the local biological diversity, natural ecosystems and landscapes for their sustenance and livelihood, with their identities being inexorably intertwined with the land they live in harmony with. Their traditions, cultural practices, diet and protection is a result of their relationship with the land, to such an extent that their lives are virtually indistinguishable from the land itself.

In many cases, these communities, having lived within the confines of their tribal structures and not being exposed to the pollution and ways of the city, are especially vulnerable to even the slightest change in their living conditions. These minor changes can threaten to disrupt not just their societal structures, but their immunity, health and general well-being.

Everyday sustainable living activities in these communities are deeply rooted in cultural and traditional behaviours, which are the foundational basis of Traditional Ecological Knowledge (TEK) practices.[38] TEK practices include hunting, gathering, fishing, and more. Disruptions in the environment and natural structures of these areas can not only disrupt these

Rank	Group	Geographic area	Documented evidence		
			Number of studies	% of total	Cumulative %
1	Inuit	Canada, Greenland, USA	31	8.18%	8.18%
2	Cree	Canada, USA	23	6.07%	14.25%
3	Ojibwe	Canada, USA	21	5.54%	19.79%
4	Dene	Canada	20	5.28%	25.07%
5	Métis	Canada, USA	15	3.96%	29.02%
6	Mohawk	Canada, USA	13	3.43%	32.45%
7	Achuar	Ecuador, Peru	10	2.64%	35.09%
8	Ogoni	Nigeria	10	2.64%	37.73%
9	Quechua	Argentina, Bolivia, Chile, Colombia, Ecuador, Peru	9	2.37%	40.11%
10	Dayak	Indonesia, Malaysia	8	2.11%	42.22%
11	Saami	Finland, Norway, Russia, Sweden	7	1.85%	44.06%
12	Maya	Belize, El Salvador, Guatemala, Honduras, Mexico	7	1.85%	45.91%
13	Navajo	USA	7	1.85%	47.76%
14	Kichwa	Colombia, Ecuador, Peru	6	1.58%	49.34%
15	Aymara	Argentina, Bolivia, Chile, Peru	5	1.32%	50.66%
16	Mapuche	Argentina, Chile	4	1.06%	51.72%
17	Sioux	Canada, USA	4	1.06%	52.77%
18	Urarina	Peru	4	1.06%	54.83%
19	Yupik	Russia, USA	4	1.06%	54.88%
20	Wayuu	Colombia, Venezuela	4	1.06%	55.94%

*Both percentages refer to the total number of studies identified in the literature review.

Table 9-1: Indigenous groups in which pollution impacts have been extensively documented.[39] Source: Integrated Environmental Assessment Management.

practices but also can also hinder the communities' ability to secure the economic gains associated with them.

In some locations[40], due to the abundance of plastic waste and its affordability, indigenous people often use it as an alternative to wood to cook their meals and keep warm. This practice is extremely toxic to their health since plastic smoke contains toxic emissions that can cause lethal effects on those who inhale them.

Far more disconcerting than the sheer scale and volume of the waste produced, however, is the longevity of the pollutants, with each individual, non-biodegradable item persisting in the natural world for decades, even centuries.

One such plastic pandemic occurred in the not-so-distant past, as a result of the global COVID-19 health crisis.

Plastic Pandemic

While the world stood still in the wake of the outbreak of the novel coronavirus at the tail end of 2019, plastic roamed free, travelling the world and wreaking havoc, while we sat safe in the comfort of our homes. From masks to sanitizer bottles, PPE suits and more, the pandemic resulted in a significant uptick in single-use medical plastics that, in most cases, were declared biohazards due to their potential to transmit the disease.

As a WHO report[41] detailed, at the peak of the pandemic 'over 140 million test kits, with a potential to generate 2,600 tonnes of non-infectious waste (mainly plastic) and 7,31,000 litres of chemical waste (equivalent to one-third of an Olympic-size swimming pool) have been shipped, while over eight billion doses of vaccine have been administered globally, producing 1,44,000 tonnes of additional waste in the form of syringes, needles and safety boxes'.

Due to the raging health crisis and uncertainties regarding person-to-person transmission, the majority of this waste was mismanaged, and ended up in landfills or waterbodies.

The chart below details the path of destruction caused by pandemic-associated mismanaged plastics to the global ocean carried via rivers.

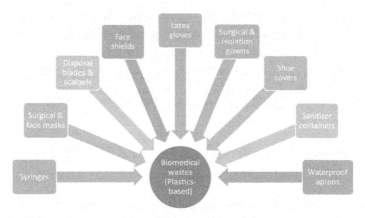

Types of plastic based biomedical wastes originated during COVID-19 pandemic.

Figure 9-15: Types of plastic-based biomedical waste that was generated during the pandemic.[42] Source: Science Direct

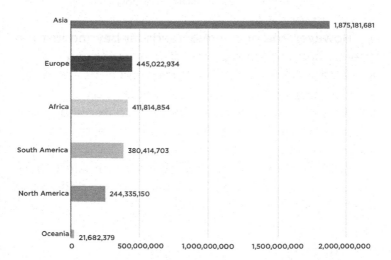

Figure 9-16: Estimated single-use face masks discarded by each continent.[43] Source: Science Direct

While for most of us, plastic waste discarded into the depths of the oceans or in fragile, isolated ecosystems does not present an immediate issue, there are others who are more directly affected, play no part in the production of the waste and cannot protest against it.

Pollution Gone Wild: The Harmful Effects of Plastic on Wildlife

Plastic pollution affects us all equally, even those of us we usually leave out of the Venn diagram of those affected. As Indians, we are no strangers to the all-too-common sight of goats, cows and more livestock consuming plastic waste from garbage dumps while searching for food.

However, the problem extends far beyond just the streets of India. From plastic straws becoming lodged in the nasal passages of sea turtles and birds and other smaller organisms becoming entangled in plastic wiring and string, our waste has dire consequences for all those who rely on the natural world for their safety and sustenance.

But how does this happen? Largely speaking, there are three key pathways by which plastic debris can affect wildlife:

Entanglement

Entanglement refers to the entrapping, encircling or constricting of animals, such as marine creatures or birds, by plastic debris.

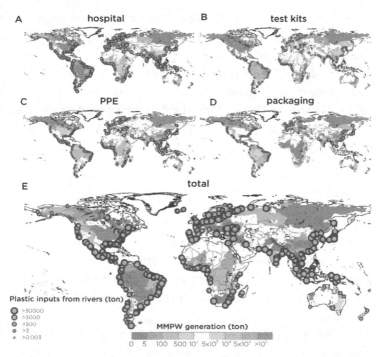

Figure 9-17: Plastic input from rivers.[44] Source:

Ingestion

As with the cows and goats of major metropolises in India, plastic can be ingested either unintentionally, intentionally, or indirectly through the ingestion of prey species containing microplastics.

The size and quantity of the plastic consumed indirectly varies on the size of the organism itself. While microplastics and nanoplastics may be consumed by fish and crustaceans, larger plastic debris, such as film, cigarette butts and more, can be found in larger fish species, and in extreme cases, even the stomachs of whales. The daily news is rife with documented cases of sperm whales ingesting very large materials including rope, plastic sheets and more.

According to Ocean Conservancy, 'Plastic has been found in 59 per cent of sea birds like albatross and pelicans, in 100 per cent of sea turtle species and more than 25 per cent of fish sampled from seafood markets around the world.'[45]

Interaction

While the term interaction seems to imply a mutual meeting of two entities, in the case of plastic debris and its harmful effects on the natural world, interaction refers to collisions, obstructions, abrasions or substrates that could cause irreparable harm to animals.

Fishing gear is a critical example of harmful interactions with plastic, causing abrasion and damage to coral reef ecosystems upon collision.

With the growing probability of plastic waste entering our oceans, it is the marine ecosystems that are most threatened by our consumption patterns. In India alone, 12.92 per cent of the waste ended up in oceans in 2019.[46]

However, it's not just fish that are the victims, as demonstrated by the following charts.

Disposal: Where Does It Go?

Now that we have discussed the evils of plastic, and its capacity for recycling, let's get to the bigger problem—

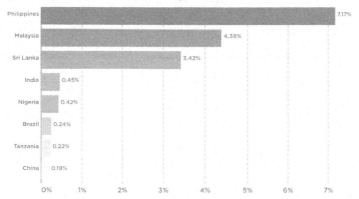

Figure 9-18: Mismanaged plastics emitted into the ocean in 2019.[47]
Source: Our World in Data

disposal. In 2015, primary plastics production was 407 million tonnes; around three-quarters (302 million tonnes) ended up as waste.[48]

This means that every time you throw a piece of plastic away, you are unwillingly contributing to one of the most pressing issues affecting our planet today.

Plastic waste majorly ends up in the ocean, with 70–80 per cent originating from the land via coastlines and rivers and 20–30 per cent from marine sources, such as fishing nets, ropes, lines and abandoned vessels. While this may be the case, 100 per cent of the waste ending up in our oceans results from the mismanagement of plastic waste.[49]

Mismanaged waste is defined as the sum of material which is either littered or inadequately disposed of, presenting a high risk of entering the ocean via wind or tides, or carried to coastlines from inland waterways.

Peer-reviewed studies demonstrating evidence ofi mpacts of plastic marine debris

Study	Animal	Encounter type	Predominant debris type	Impact (response)
Allen et al. 2012	Grey seals	Entanglement	MF line, net, rope	Constriction
Beck & Barros 1991	Manatees	Entanglement	MF line, bags, other debris	Death
Campagna et al. 2007	Elephant seals	Entanglement	MF line, fishing jigs	Dermal wound
Croxall et al. 1990	Fur seals	Entanglement	Packing bands, fishing gear, other debris	Dermal wound
Dau et al. 2009	Seabirds, pinnipeds	Entanglement	Fishing gear	External wound
Fowler 1987	Fur seals	Entanglement	Trawl netting, packing bands	Death
Fowler 1987	Fur seals	Entanglement	Trawl netting, packing bands	Reduced population size
Good et al. 2010	Invertebrates, fish, seabirds, marine mammals	Entanglement	Derelict gillnets	Death
Moore et al. 2009	Seabirds marine mammals	Entanglement	Plastic, fishing line	Death
Pham et al. 2013	Gorgonians	Entanglement	Fishing line	Damage/breakage
Mez-Rubio et al. 2013	Sea turtles	Entanglement	Fishing gear	Death
Winn et al. 2008	Whales	Entanglement	Plastic line	Dermal wound
Woodward et al. 2006	Whales	Entanglement	Plastic line	Dermal wound
Beck & Barros 1991	Manatees	Ingestion	MF line, bags, other debris	Death
Bjorndal et al. 1994	Sea turtles	Ingestion	MF line, fish hooks, other debris	Intestinal blockage, death
Boesch. 2011	Penguins	Ingestion	Plastic, fishing gear, other debris	Perforated gut, death
Browne et al. 2013	Lugworms (laboratory)	Ingestion	Microplastics	Biochemical/cellular, death
Bugoni et al. 2001	Sea turtles	Ingestion	Plastic bags, ropes	Gut obstruction, death
Carey 2011	Seabirds	Ingestion	Plastic particles, pellets	Perforated gut
Cedervall et al. 2012	Fish (laboratory)	Ingestion	Nanoparticles	Biochemical/cellular
Connors & Smith 1982	Seabirds	Ingestion	Plastic pellets, foam	Biochemical/cellular
Dau et al. 2009	Seabirds, pinnipeds	Ingestion	Fishing hooks	Internal wound
de Stephanis et al. 2013	Sperm whale	Ingestion	Identifiable litter items	Gastric rupture, death
Fry et al. 1987	Seabirds	Ingestion	Plastic fragments, pellets, identifiable litter	Gut impaction, ulcerative lesions
Jacobsen et al. 2010	Sperm whales	Ingestion	Fishing gear, other debris	Gastric rupture, gut impaction, death
Lee et al. 2013	Copepods (laboratory)	Ingestion	Micro- and nanoplastics	Death
Oliveira et al. 2013	Fish (laboratory)	Ingestion	Microplastics	Biochemical/cellular

Table 9-2: Plastics in marine environment.[50] Source: Research Gate

The famous Great Pacific Garbage Patch (GPGP)[51] is a great example of the prevalence and persistence of plastic waste in our natural environment.

A patch of floating plastic waste formed by natural ocean gyres, that is, circular ocean currents brought about by global wind patterns, the GPGP is a collection of marine debris in the North Pacific Ocean. Fifty-two per cent of it is composed of plastic waste from marine

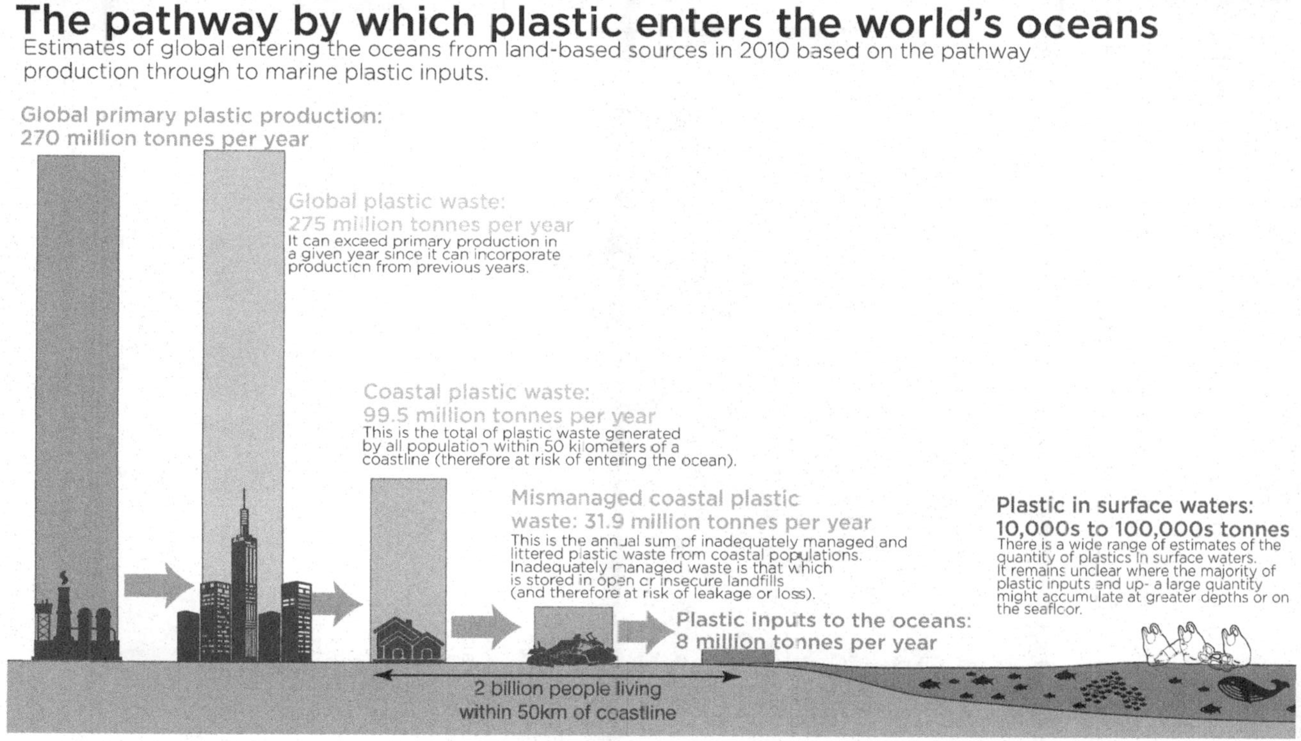

The pathway by which plastic enters the world's oceans

Estimates of global entering the oceans from land-based sources in 2010 based on the pathway production through to marine plastic inputs.

Global primary plastic production: 270 million tonnes per year

Global plastic waste: 275 million tonnes per year
It can exceed primary production in a given year since it can incorporate production from previous years.

Coastal plastic waste: 99.5 million tonnes per year
This is the total of plastic waste generated by all population within 50 kilometers of a coastline (therefore at risk of entering the ocean).

Mismanaged coastal plastic waste: 31.9 million tonnes per year
This is the annual sum of inadequately managed and littered plastic waste from coastal populations. Inadequately managed waste is that which is stored in open or insecure landfills (and therefore at risk of leakage or loss).

2 billion people living within 50km of coastline

Plastic inputs to the oceans: 8 million tonnes per year

Plastic in surface waters: 10,000s to 100,000s tonnes
There is a wide range of estimates of the quantity of plastics in surface waters. It remains unclear where the majority of plastic inputs and up- a large quantity might accumulate at greater depths or on the seafloor.

Figure 9-19: Ways in which plastics enter the oceans.[52] Source: Our World in Data

sources due to intensive fishing activity in the Pacific Ocean.

A compounded entity formed by the collision and merging of the western (Japan) and eastern (USA)

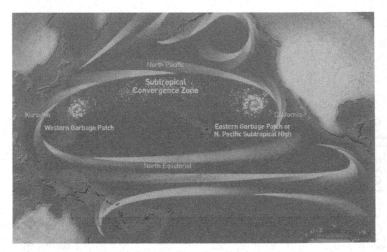

Figure 9-20: The Great Garbage Pacific Patch.[53] *Source: National Geographic*

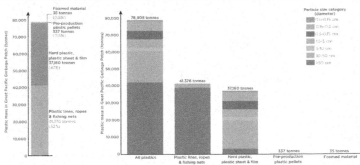

Figure 9-21: The components of the great pacific garbage patch.[54] *Source: Our World in Data*

garbage patches interconnected by the subtropical convergence zone, The GPGP is responsible for the deaths of one million seabirds and 1,00,000 marine mammals every year, as well as many other species.[55]

A sampling of the material in the GPGP showed that 94 per cent of the 1.8 million pieces of plastic are microplastics accounting for just 8 per cent of the plastic by weight. Most of the plastic by weight were larger pieces of plastic, with fishing nets accounting for 46 per cent.[56]

What Are We Missing?

Our global struggle against plastic is, to me, just another example of humanity's greed and our quest for convenience, backfiring. Plastic is not evil, but rather is a tool wielded by evil people who would rather fill their pockets and watch the world burn than put a stop to the damage.

What we all need to understand is that, when the plastic industry benefits, so does the fossil fuel industry. Produced via by-products of coal and oil refinement, (which would have been expensive to offset) fossil fuel companies found a way to turn their waste into gold with little cost to their operations.

Think about it like this—plastic is perhaps one of the most commonly consumed, traded and sold materials on the planet. If you had a paisa for every piece of plastic ever produced in the world, how rich would you be?

This money is not an intangible metric but constitutes a larger part of the reason why global society has had such a difficult time going zero-plastic.

The gatekeepers of the industry are the ones holding the strings and you and I are mere pawns in their stage acting.

While premium brands overprice their products to mirror their brand value, plastic products are not branded and are marketed to be cheap, replicable and disposable. This is not mere happenstance but is a deliberate move to mint money, no matter the consequences.

Another important factor to consider is that currently we don't have a suitable alternative to plastic, that can withstand impact, contain liquids without spilling, and is government certified as safe to use. But does that mean such a solution does not exist?

Of course not.

However, with such a cheap alternative on the market, governing bodies and manufacturers are not properly incentivised to find a solution.

But what about us? Are we not properly incentivised?

When you buy a bottle of water, you are paying for the water, because you are thirsty. Unbeknownst to you, you are also being charged a small tax towards the destruction of your environment in the form of the bottle in which it is sold. When you buy a packet of biscuits, you are paying for your hunger with your money. However, you are also paying for the plastic wrapper, as well as the tray within, which keeps your biscuits from crumbling during transportation.

With every rupee spent, a certain amount goes towards the destruction of your planet—to me, this seems a wasteful practice to spend my hard-earned money on.

When I turned vegan, I realised very quickly that plastic means the destruction of not just the health of my planet, but my body and mind as well. Today, plastic packaging is not made to contain food, but rather food is pumped full of preservatives and unhealthy trans fats to suit its packaging. This is the extent to which plastic rules all of us.

Of the 5800 million tonnes of primary plastic no longer in use, only 9 per cent has been recycled since 1950.[57] While it is easy to say that government policies on recycling and plastic use are too lax, what is harder to admit is that the government works for us—if we do not demand change, change will never come.

Before major retail stores began charging just Rs 1 for a plastic bag along with your groceries, you never blinked an eye while watching the clerk pack your purchases into thin plastic bags. However, the introduction of the plastic tax, as I like to call it, has seen a shift in consumer mindsets, with more and more people hesitating to pay an additional rupee, even though they have just swiped their cards for thousands of rupees of groceries.

The simple fact of the matter is, we don't respect what we get for nothing until it threatens to destroy us all.

So, what do we do about it?

Solutions: (Pla)sticking it to the Man

While the movers and shakers behind the scenes may not bat an eye in the fight for zero-plastic, we consumers can make great strides towards changing the face of our relationship with unsustainability.

The simplest way to do this is to focus on the 'four Rs' of plastic waste management—Refuse, Reuse, Reduce and Recycle.

According to Gunjan Jindal Poddar, founder of Amala Earth:

'I believe that brands must take the onus of encouraging ethical and eco-sensitive consumerism in order to facilitate change in society at a macro level. Sustainability is a mere stepping stone or a facilitator in the process of imbibing and imbuing conscious transformation towards earth-friendliness.'

Refuse

The power of saying no is a highly underrated skill. It has the power to save countless lives. Whenever possible, limit your use of plastic products.

Reuse

Just because plastic waste is unsustainable, doesn't mean our use of it should be. As far as possible, reuse your plastic products. From plastic bags to certain containers, plastic products can be reused to a large extent.

However, certain types of plastic can leach harmful chemicals into the contents it holds and can even

Figure 9-22: Why you should refuse single-use plastic.[58] Source: Her Planet Earth

deliver microplastics into the contents via weathering and friction.

Do your research and stay safe, while keeping the safety of our environment in mind as well.

Reduce

While eliminating plastic from our daily lives can indeed seem like an impossible undertaking, even a small reduction in our consumption can save our planet from centuries of destruction.

Nandan Bhat, founder and director of EcoKaari, says:

> 'Indian plastic consumption numbers are some of the highest anywhere in the world, and this presents an economic opportunity for us to become one of the world's biggest recycling markets. Brands are bringing together zero-waste design, small-scale production, time-tested Indian traditions, centuries-old recycling practices, and up-cycling to cater to consumers. Upcycled, ethically-made, eco-friendly, reusable products are the way forward.'

Becoming a plastic-free consumer is a noble pursuit, but until more and more manufacturers begin to affect real change, it will be a difficult task for even the most environmentally conscious city-dweller. However, the shift is already occurring on an industrial level, and can be catalysed by small changes to our everyday behaviours.

For example, when shopping, carry a reusable bag to store your groceries and other items. This will not only reduce the plastic waste being generated and abandoned to the natural world, but also will also keep the contents of your shopping bag safe from chemical pollutants.

As far as possible, limit your use of single-use plastic and plastic products.

Recycle

Recycling may still be in its nascent stages as a method for the disposal of plastic, but we are getting better at it with every passing year.

This is evident from the progression of recycling itself. Before 1980, recycling and incineration of plastic was negligible; 100 per cent was therefore discarded.

Time			1990	2000	2010	2019
Location						
OECD	OECD America	United States	0.821	2.248	4.342	6.455
		Canada	0.076	0.258	0.543	0.84
		Other OECD America	0.14	0.452	0.995	1.684
	OECD Europe	OECD EU	0.183	3.098	7.832	12.767
		OECD Non-EU	0.024	0.442	1.239	2.292
	OECD Pacific	OECD Asia	0.022	0.43	1.314	2.592
		OECD Oceania	0.004	0.025	0.09	0.217
Non-OECD	Other America	Latin America	0.092	0.471	1.466	2.826
	Eurasia	Other EU	0.002	0.046	0.131	0.249
		Other Eurasia	0.07	0.268	0.717	1.362
	Middle East & Africa	Middle East & North Africa	0.081	0.291	0.727	1.383
		Other Africa	0.044	0.225	0.692	1.328
	Other Asia	Other non-OECD Asia	0.061	0.421	1.533	3.693
	China		0.036	1.302	5.868	13.23
	India		0.006	0.259	1.266	3.66

Table 9-3: Plastic waste collected for recycling by countries. Source: OECD[59]

From 1980 for incineration and 1990 for recycling, rates increased on average by about 0.7 per cent per year. In 2015, an estimated 55 per cent of global plastic waste was discarded, 25 per cent was incinerated, and 20 per cent was recycled.

If your area is not equipped with a recycling facility, appeal to your local government to set up a recycling programme. Encouraging community participation towards sustainability is a great way to educate people, while also making a change for the better.

Sahar Mansoor, co-founder of Bare Necessities, says:

'Solid waste management is one of the greatest costs to municipal budgets. However, we are currently seeing an increase in more consumers looking to purchase conscious brands, a pattern that is particularly pronounced among young consumers, who actively choose to prioritize sustainability and ethical values when making purchasing decisions. Generation Z, in particular, is becoming more mindful of and asking the right questions: 'What's in my products?', 'Who made my clothes?' 'How much did they earn in the process?'. The emerging conscious millennial population wants to align itself to certain causes by virtue of its consumption choices. And this is then percolating down to further generations, while also having an impact on the older generations.'

Greening Personal Care

When it comes to limiting plastic waste, we must start from the source, that is, ourselves. Here are a few ways you can choose to go plastic free with your personal care.

Switch to Metal or Wood

While we may not actively think about it, disposable plastic razors, comb, and more can become an environmental hazard rather quickly. Instead, invest in a longer-term solution such as wooden or metal products. While these may be marginally more expensive than their single-use plastic counterparts, they are proportionately more durable, and will save you money in the long run.

From metal razors to wooden combs, you would be surprised at how much plastic can be easily eliminated from your personal care routine.

Dental Hygiene

While toothpaste is generally considered a necessity, it contributes to plastic waste via its packaging. Consider switching to tooth powder or tablets, which are generally stored in eco-friendly packaging and get the job done just as well as a regular tube of paste.

Another single-use plastic product that most tend to overlook is floss. While indeed minuscule in size, floss is technically plastic waste, and with over seven billion flossers around the world, even one small string can quickly turn into a ball of yarn that is difficult to unravel. Switch to silk dental floss, or a Waterpik to make your dental hygiene routine more eco-friendly![60]

Menstrual Routines

As we have discussed previously, sanitary napkins, panty liners and tampons are either made using plastic or lined with it. Adding to the problem, they also often come wrapped individually in thin plastic.

While feminine hygiene products are indeed a necessity, there has been a wave of new start-ups investing in green solutions to menstrual hygiene.

The growing demand for sustainable alternatives to feminine hygiene has seen several homegrown brands investing in multiple-use alternatives with innovations such as period underwear, reusable fabric pads, and menstrual cups promising to change the face of plastic waste from menstrual routines overall.

Boondh is one such company that is making great strides not just toward the eradication of single-use plastic feminine hygiene products, but also towards educating rural women on the importance of sustainability and personal hygiene.

Green Dining

In our fast-paced lives, we often must find time to grab a quick bite between demanding tasks, such as work, commuting, and more. However, this lifestyle is a significant contributor to the plastic problem. As mentioned in Chapter 4, travel kits can go a long way towards eliminating plastic waste while dining on the go. Equipped with reusable cutlery, straws and plates, a travel dining kit is the quickest solution to greening on the go.

Plastic pollution is perhaps the single-biggest issue plaguing our world today, and only we can do something about it. I invite you to pledge to a plastic-free life, one that is kinder to our environment, our health, and that of countless others who depend on our natural world for their survival.

Resources

Here is an inventory of supplementary media material, people who matter and shops in line with the chapter's theme.

Books

1. *No More Plastics: What You Can Do to Make a Difference*
   Author: Martin Dorey
   Published by: Ebury Press
   https://www.amazon.in/No-More-Plastic-difference-2minutesolution/dp/1785039873

2. *Turning The Tide on Plastic: How Humanity (and you) Can Make Our Globe Clean Again.*
   Author: Lucy Siegle
   Published by: Trapeze
   https://www.amazon.in/Turning-Tide-Plastic-Humanity-Globe/dp/1409182983

3. *The Climate Solution*
   Author: Mridula Ramesh
   Published by: Hachette India
   https://www.amazon.in/Climate-Solution-Indias-Climate-Change-Crisis/dp/9388322207

4. *The Plastic Problem: 60 Small Ways to Reduce Waste and Help Save the Earth (Kids)*
   Author: Aubre Andrus
   Published by: Lonely Planet Global

https://shop.lonelyplanet.com/products/the-plastic-problem-1-row

5. *How Bad are Bananas? Carbon Footprint of Everything*
   Author: Mike Berners-Lee
   Published by: Profile Books
   https://www.amazon.in/How-Bad-Are-Bananas-everything/dp/1846688914

Documentaries

1. *A Plastic Ocean*
   Directed by: Craig Leeson
   Available on Amazon
   https://www.aplasticocean.movie

2. *The Story of Plastic*
   Directed by: Deia Schlosberg
   Available on Amazon
   https://www.storyofstuff.org/movies/the-story-of-plastic-documentary-film

3. *How People Profit Off India's Garbage | World Wide Waste Series*
   Directed by: Business Insider Team
   Available on Youtube
   https://www.youtube.com/watch?v=V7TcEnSOR3s

Ted Talks

1. Going The Green Way
   By: Meenakshi Bharath

https://www.youtube.com/watch?v=AzqG
1U9ekNc

2. How I turned Plastic Problem into a Housing
    Solution
    By: Rushabh Chheda
    https://www.youtube.com/watch?v=iVy4
    FcA_p3Y

3. Turning Plastic Back to Fuel
    By: Dr. Medha Tadpatrikar
    https://www.youtube.com/watch?v=YrjW
    dz0LEM0

4. Bamboo's Boon: An Entrepreneur's Mission to
    Reduce Plastic Waste
    By: Yogesh Shinde
    https://www.youtube.com/watch?v=mEV8Hq-
    Q9YA&t=433s

5. How to Reduce Your Waste
    By: Nila Patty
    https://www.youtube.com/watch?v=1TDC-Zud_uM

People

1. The Minimalists | @theminimalists
    https://www.instagram.com/
    theminimalists/?hl=en

2. Mrudula | @Ullisu.official
    https://www.instagram.com/ullisu.official

3. Megha Pandey | @nativekeeps
    https://www.instagram.com/nativekeeps

4. Malvika | @malvika_r
    https://www.instagram.com/malvika_r

Shop

1. Brown Living
    Delivers PAN India
    https://brownliving.in

2. Bare Necessities
    Delivers PAN India
    https://barenecessities.in/

3. Almitra Sustainables
    Store in Mumbai | Delivers across India
    https://www.almitrasustainables.com

4. Recharkha
    Upcycled Bags made from Plastic
    https://www.recharkha.org

5. Rescript Stationery
    Sustainable/Recycled Stationery
    https://www.rescript.in

6. Thooshan
    Edible Cutlery and Tableware
    https://thooshan.com

7. Mitticool
   Eco-Friendly Clay Products
   https://mitticool.com

Donate here

1. The Kabadiwala
   Sell Your Scrap
   https://www.thekabadiwala.com

2. Ecokaari
   Donate Plastic
   https://www.ecokaari.org/pages/donate-plastic-to-ecokaari

10

Our Beliefs and Choices

This industry impacts the following Sustainable Development Goals (SDGs):
- SDG 3: Good Health and Well-Being
- SDG 6: Clean Water and Sanitization
- SDG 9: Industrial Innovation and Infrastructure
- SDG 12: Responsible Consumption and Production
- SDG 13: Climate Action
- SDG 14: Life Below Water
- SDG 15: Life on Land
- SDG 16: Peace, Justice and Strong Institutions
- SDG 17: Partnership for the Goals

ACTIVITY

Before we begin this chapter, try and answer the following questions as honestly and thoroughly as possible:

1. What faith do you subscribe to?
    Please specify: _____

2. Do you actively participate in your faith?
    ☐   Yes
    ☐   No
    ☐   Sometimes

3. What do/did you study?
    Please specify: _____

4. What will/do you do for work?
    Please specify: _____

5. Do you invest your money?
    ☐   Yes
    ☐   No

Every day, we are faced with a multitude of choices. While some choices are mundane, like what to eat for dinner, others can present a more pressing challenge.

From what we eat to what we wear, and from what we study to what we do for money — every single minute of our collective existence is a reflection of over seven billion people making a choice.

But what drives these choices?

No matter where or how you grew up, our preferences, beliefs and behaviours reflect our childhood experiences. From the food we eat to the music we listen to, the languages we speak and our religious and political temperaments—our social and familial structures play a big role in our development.

But what if I were to tell you that the opposite is also possible?

While it is true that our mature, adult selves are amalgamations of our infant and adolescent experiences, moulded by our families and the society within which we grew up? Our beliefs and teachings today can change the way the next generation learns, grows, believes and develops.

Belief is a potent tool and can be used for good or evil. Belief drives all that we do, from our everyday interactions with one another to our culture, traditions, and even the way our countries and nations are governed.

However, the pioneers and architects of modern society have found a way to exploit belief to turn a profit, perverting what was once pure, functional and equal until we are left with nothing but another capitalistic structure.

If we are both simultaneously products and sculptors of our societal beliefs, we can further infer that we are both victims and perpetrators of the ills of our modern world.

In a world overrun by conflicting belief systems, can belief itself be a path to unity? If every man, woman and child living on our planet today were able to unite under the same umbrella of belief, how would that change the way we live?

Let's find out.

Penny for Your Thoughts: The Problem with Beliefs

When I say the word 'belief', what is the first thing that pops into your mind?

While most associate the term with purely religious beliefs, every single choice we have ever made, or will make, is guided by our personal set of beliefs.

If you believe in making the world a better place by healing people, chances are you will choose to study biology, chemistry, physics and medicine, before choosing to become a doctor, surgeon, or medical practitioner.

If you are a child with a fascination for space, chances are you will probably pursue this passion, studying the cosmos and unravelling the mysteries of our universe. As you learn, perhaps that initial fascination transforms into a belief, one that states that the final frontier presents humanity's only hope for survival. Guided by this belief, you may choose to enrol in an astrophysics course, eventually applying for a position a job that involves designing and manufacturing components for extra-terrestrial aeronautics.

If you believe that you have gained too much weight over the festive season, chances are you will choose a fad diet recommended by your friend. If the diet worked for you, you will likely recommend this diet to another friend, who may choose to try it or not.

If you believe that this political party or the next can change the world and bring equality to your people, whose own beliefs have been side-lined for generations, you will likely choose to go out on election day to cast your vote for it.

If you believe the only salvation for your sins is piety and devotion to your God or gods, chances are you will choose to say a silent prayer for forgiveness, redemption and kindness, before you close your eyes to sleep for the night.

At every turn, with every choice, our beliefs serve to caution, motivate and guide us.

However, beliefs are not set in stone and can change from one moment to the next. My hope is that, through this narrative, I can change some of your beliefs and thereby change the world that you and I live in.

However, there is an obstacle to this undertaking— one that would rather keep us stuck in a vicious cycle of outdated beliefs and social prejudices.

What is this obstacle, you ask?

The answer is money or rather, the societal structures built around the flow of money from the majority to the minority—capitalism.

Let's take a closer look at how capitalism has infiltrated our beliefs, perverting them into unstoppable machines of unsustainability, inequality and destruction.

Mera Bharat Mahaan: A Closer Look at India

India is a nation of diverse beliefs, traditions and cultures. A melting pot of languages, ethnicities, histories and knowledge, India is home to at least six major religious groups, with countless other smaller groups and communities.

India speaks in a multitude of languages and has just as many voices to match.

As per the 2011 census, 43.63 per cent Indians said Hindi is their mother tongue. It was followed by Bengali with 8.03 per cent speakers, Marathi with 6.86 per cent, Telugu with 6.7 per cent and Tamil with 5.7 per cent.

Thirteen of the twenty-two scheduled languages were reported as the mother tongue by at least 1 per cent of the population.

While language and religion have been the two major causes of conflict among social groups across the world, India has existed in a state of harmony, a commendable accomplishment for a nation of over 138 crore people.[1]

However, recent times have seen an uptick in communal violence, religious differences, caste wars

ADMINISTRATIVE DIVISIONS 2011

Sl. No.	India/States/Union Territories	Number of Districts	Area (Km²)#	Population '@'	Major Languages
s	t	u	v	w	x
	INDIA	640	32,87,469	1,21,08,54,977	Hindi, Bengali, Marathi
1	JAMMU & KASHMIR	22	2,22,236	1,25,41,302	Kashmiri, Hindi, Dogri
2	HIMACHAL PRADESH	12	55,673	68,64,602	Hindi
3	PUNJAB	20	50,362	2,77,43,338	Punjabi
4	CHANDIGARH	1	114	10,55,450	Hindi
5	UTTARAKHAND	13	53,483	1,00,86,292	Hindi
6	HARYANA	21	44,212	2,53,51,462	Hindi
7	NCT OF DELHI	9	1,483	1,67,87,941	Hindi
8	RAJASTHAN	33	3,42,239	6,85,48,437	Hindi
9	UTTAR PRADESH	71	2,40,928	19,98,12,341	Hindi
10	BIHAR	38	94,163	10,40,99,452	Hindi, Marathi
11	SIKKIM	4	7,096	6,10,577	Nepali
12	ARUNACHAL PRADESH	16	83,743	13,83,727	Nissi/Dafla, Adi
13	NAGALAND	11	16,579	19,78,502	Konyak, Ao, Sema
14	MANIPUR	9	22,327	28,55,794	Manipuri
15	MIZORAM	8	21,081	10,97,206	Lushai/Mizo
16	TRIPURA	4	10,486	36,73,917	Bengali
17	MEGHALAYA	7	22,429	29,66,889	Khasi, Garo

Sl. No.	India/States/Union Territories	Number of Districts	Area (Km²)#	Population '@'	Major Languages
s	t	u	v	w	x
18	ASSAM	27	78,438	3,12,05,276	Assamese, Bodo
19	WEST BENGAL	19	88,752	9,12,76,115	Bengali
20	JHARKHAND	24	79,716	3,29,88,134	Hindi
21	ODISHA	30	1,55,707	4,19,74,218	Odia
22	CHHATTISGARH	18	1,35,192	2,55,45,198	Hindi
23	MADHYA PRADESH	50	3,08,252	7,26,26,809	Hindi
24	GUJARAT	26	1,96,244	6,04,39,692	Gujarati
25	DAMAN & DIU	2	111	2,43,247	Gujarati, Hindi
26	DADRA & NAGAR HAVELI	1	491	3,43,709	Bhili/Bhilodi, Hindi
27	MAHARASHTRA	35	3,07,713	11,23,74,333	Marathi
28	ANDHRA PRADESH	23	2,75,045	8,45,80,777	Telugu
29	KARNATAKA	30	1,91,791	6,10,95,297	Kannada
30	GOA	2	3,702	14,58,545	Konkani
31	LAKSHADWEEP	1	30	64,473	Malayalam
32	KERALA	14	38,852	3,34,06,061	Malayalam
33	TAMIL NADU	32	1,30,060	7,21,47,030	Tamil
34	PUDUCHERRY	4	490	12,47,953	Tamil
35	ANDAMAN & NICOBAR IS.	3	8,249	3,80,581	Bengali, Hindi, Tamil

Table 10-1: Languages spoken across various states in India.[2] Source: Census of India

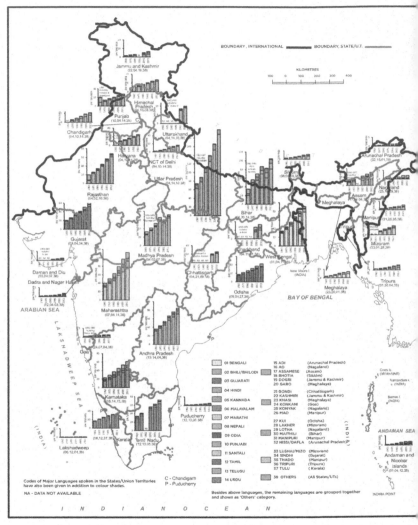

Figure 10-1: Variation in speakers of major languages.[3] Source: Census of India

and political instability dominating headlines all over the nation. But there is another, silent war raging just beneath the surface, one that weaponizes belief for personal gain, pitting us against one another rather than uniting us towards a brighter future for our planet and our people.

Devotion and Devastation: The Perversion of Tradition

India is home to several religious groups, each with its own set of beliefs, customs and traditions. Hinduism has the largest number of followers.

While India's secularism is perhaps its most discussed, and yet precious virtue, our freedom to practice our religious customs, at times, whether rightly or wrongly comes with many hidden costs to our environment, people and all living things.

With thousands upon thousands of religious festivals and observances occurring every year, the nation as a whole operates in an almost constant state of celebration, with regional festivities marked by lavish feasts, parades, the exchange of gifts, the lighting of electric lights and the bursting of firecrackers.

However, as with most religious gatherings, sites of communal get-togethers are strewn with litter and solid waste in the form of single-use plastic (SUP) plates, cutlery, beverage glasses and more, most of which is mismanaged. In general, rampant food waste generation, over-consumption of electricity and water, and noise, air, water and land pollution associated with traditional practices become the norm.

The demand for beef, mutton, chicken and other meat products during some of these occasions has been known to increase by around 50 per cent of the usual demand. Adding to the problem, much of this meat is imported from foreign nations, adding to the environmental cost.

Religious festivals are also a time for family, meaning that thousands of people from across the globe will travel far and wide to visit their loved ones, adding to the carbon footprint.

As we have already discussed, flying is perhaps one of the most environmentally damaging travel behaviours, and with the concentration of fliers increasing around the festive season, the carbon footprint of each traveller increases tenfold.

The most puzzling factor in the corruption of our traditions, however, is that all religions are rooted in Paganism, that is, the practice of worshipping the natural world and the forces of nature. However, in the modern world, observances rooted in the celebration and veneration of nature are the cause of environmental destruction.

While this duality is especially prevalent in India, the impact of consumerism on our faith is a universal ordeal, with capitalistic institutions benefitting greatly from the exploitation of cultural and traditional practices—whether it is the firecracker industry employing child labourers to create delicate combustible products using harmful chemicals and toxic metals, or the many decorations, gifts and inorganic offerings that go to waste in the form of particulate matter, water-borne sediment, solid waste and more.

But what if I were to tell you that capitalism has many festivals of its own?

Large-Scale Capitalistic Festivals

Did you know that a year of retail comprises six major seasonal sales? If you, like me, live in India, these seasonal sales are also divided into many smaller sales for festivals, sports events, clearances and more.

In every sale cycle, retailers and service providers globally sell thousands of products, many of which are single-use or frivolous. While a little retail therapy never hurt anyone, and the availability of sales every other month does make it easier for shoppers on a tight budget to avail of the goods and services they seek, certain observances are designed to be purely capitalistic, hidden beneath the guise of a cultural observance.

Greenmyna founder Ashwin Malwade says:

'For centuries, our country's religious and cultural events were sustainable and in sync with nature. It is only the last two decades that have seen our religious and cultural events take on a capitalist tone, where the narrative has been to celebrate events on a grand scale with no regard for nature. Making an event sustainable is not a shift from the normal, it's what our ancestors had been doing for generations. India is a land of a thousand events and large-scale celebrations, whether personal, religious, national or international. In a nation that loves celebrations, and spares no expense to make them as memorable as possible, there is a large scope for sustainable players in the market. After all, the health of our environment is not a cost we should take lightly.'

Thanksgiving

While not a religious observance, Thanksgiving is a national holiday in the United States and Canada that has roots in a pagan harvest festival.

According to a study by the University of Manchester, an average Thanksgiving dinner for eight guests contributes approximately 44 pounds of carbon dioxide emissions into the atmosphere, with the transportation of the meat, and the meat itself being associated with most of this number.[4]

While Thanksgiving is indeed a national observance, modern traditions have given rise to two associated capitalistic festivals, namely Black Friday and Cyber Monday, that have since crossed the vast ocean to make their way into many other countries, including India.

According to a 2018 estimate, American consumers generated an average shopping bill of $1007.24, with much of the spending occurring on Black Friday and Cyber Monday.[5]

Black Friday occurs on the Friday after Thanksgiving, marking the beginning of the holiday shopping season. With unbelievable discounts and limited stock, Black Friday sales have seen people clamouring over one another to get to the last piece of their desired goods. Seventeen deaths and 125 injuries

PAKISTAN	11525%	SPAIN	646%
HUNGARY	9750%	INDIA	822%
ITALY	4516%	PORTUGAL	607%
GREECE	2600%	CHILE	564%
RSA	2571%	BULGARIA	562%
BELARUS	2279%	IRAN	531%
GERMANY	2106%	SWEDEN	530%
USA	2103%	RUSSIA	527%
UAE	1913%	KENYA	492%
AUSTRALIA	1764%	HONG KONG	444%
CANADA	1732%	QATAR	391%
KAZAKHSTAN	1675%	SLOVAKIA	387%
FINLAND	1529%	CZECHIA	355%
UNITED KINGDOM	1500%	MOROCCO	353%
NIGERIA	1342%	ARGENTINA	325%
UKRAINE	1303%	SINGAPORE	285%
ROMANIA	1260%	NEW ZEALAND	261%
SAUDI ARABIA	1250%	KUWAIT	253%
IRELAND	1231%	PUERTO RICO	214%
AUSTRIA	1202%	MEXICO	179%
BRAZIL	1094%	ECUADOR	156%
EGYPT	940%	COLOMBIA	154%
SWITXERLAND	934%	VIETNAM	145%
INDONESIA	849%	DOMINIKAN REP.	145%
PHILIPPINES	740%	VENEZUELA	119%
TURKEY	697%	PERU	107%
POLAND	692%	THAILAND	46%
MALAYSIA	651%	GLOBAL AVERAGE	624%

Figure 10-2: How do sales go on Black Friday compared to a regular Friday?[6] Source: Black Friday Global

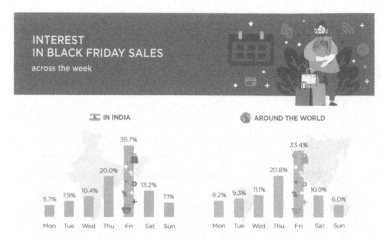

Figure 10-3: *Interest in Black Friday sales in India and World.*[7] *Source: Black Friday Global.*

have been recorded during this sae around the world.[8] While some of this number is associated with armed or intentional violence, a significant portion represents people caught in stampedes as stores open. An average Indian will spend over Rs 4500 during Black Friday sales, with clothes, electronics and shoes being the most popular purchases.[9]

Cyber Monday, much like its predecessor, is a capitalistic festival associated with online shopping. It has major discounts on electronic goods, as well as items on aggregator retail websites.

The Big Fat Indian Wedding

While religious and capitalistic festivals come with associated environmental and social footprints, cultural observances, such as weddings and other celebrations, are not far behind.

One pressing example of the performative, capitalistic perversion of a cultural event is the Big Fat Indian Wedding—named so due to its over-the-top, no-expense-spared nature.

Regardless of what religion one follows, no one does a wedding like Indians. With a concentration of SUP cutlery, crockery and more, as well as solid waste in the form of floral installations, décor, return gifts and more, the carbon footprint of a typical Indian wedding is the equivalent of approximately 100 trees being felled.

Food is by far the biggest source of waste generated by weddings. On an average 3000-4000 kg is wasted with each celebration. In Bengaluru alone, an average of 85,000 weddings occur annually, throwing away a cumulative 943 metric tons of food.[10]

While I am not saying that religious festivals, traditional customs, the buying of gifts or the celebration of love and the unison of families is bad, or that the people that participate in them are knowingly harming the planet, or that the religions are wrong in any way—yet I would truly like to draw your attention to the choices we make in the name of family, love, respect, tradition and faith.

Capitalistic structures like the one we live in today are nothing more than a cluster of elaborate cons, making us believe that in order to truly 'participate', money must pass from one segment of society to another. They also dupe us into forgetting the true meaning of these observances, which, in most cases is a form of tribute to nature and the power of the universe. Instead of commemorating these virtues of the world, as these festivals and observances were meant to do, modern society has made nature itself the victim of a cruel joke with the fate of humankind being the punchline.

From packaging to gratuitous buying, these celebrations have been twisted and perverted into money-making businesses that have turned these commemorations of life into causes of death.

However, humans are free-thinking, sentient beings with the ability to choose. It is our choices that brought us to this place, and it is our choices alone that can save us, and the traditions and customs we hold so dearly in our hearts.

We must choose to bridge the gap between belief and practice and stay true to the meaning of these symbolic acts rather than be swayed by the shiny bits and baubles of communal and capitalistic activities.

'But Aakash', I hear you ask, 'how are we to know which choice is the right one for our cultural heritage, familial and societal structures and environmental health?'

The answer is simple—make educated decisions. Explore your options and choose wisely.

But what about the choices we make towards educating ourselves?

By the Book: The Problem with Educational Institutions

For most of us, school is where our lives truly begin. It is where we first learn to think independently and meet and interact with peers and mentors independent of our familial structures. School is where our beliefs first take shape.

From math to biology, hygiene and social skills, educational institutions offer us our first taste of the real world, where we may make any number of choices guided by our learnings.

India has long been plagued by dismal literacy rates, with a 2021 estimate standing at 74.04 per cent. While this number does seem significant and is a 12 per cent increase from our literacy rates when we first secured Independence, India is still behind the global average literacy rate of 84 per cent.

While this disparity between the literate and illiterate population of India is indeed a cause for concern, an even greater concern is the quality of education most students receive.

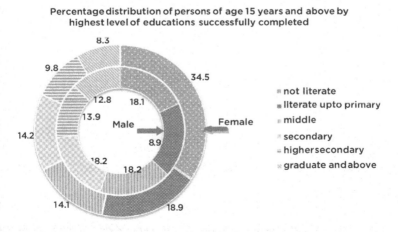

Percentage distribution of persons of age 15 years and above by highest level of educations successfully completed

Male — Female

not literate
literate upto primary
middle
secondary
higher secondary
graduate and above

Figure 10-4: Percentage of persons above the age of 15 years who have completed higher education.[11] *Source: Government of India.*

As of 2018, the number of students enrolled in government schools and institutions in India stood at 131 million, and 119 million enrolled students were enrolled in private schools.[12]

However, as demonstrated by the chart, government expenditure in education has been on a steady decline since 2014. These budget cuts mean fewer or worse amenities for students, fewer optional curricular choices, lower salaries for teachers and staff and associated lower quality of upkeep and hygiene in these institutions.

According to nationwide averages, a government school teacher typically earns a meagre salary of

Spending On Education Down Since 2014

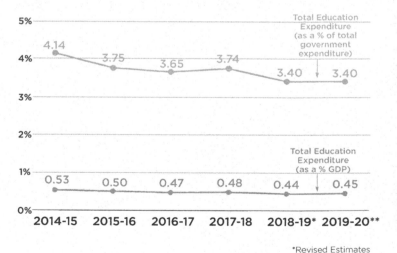

Figure 10-5: Spending on education down since 2014.[13] Source: India Spend

Rs 25,000 per month, with a requirement to work a minimum of 45 hours a week.[14]

With this in mind, it comes as no surprise that the more qualified and experienced teachers would choose to teach elsewhere, where their time and effort would be sufficiently rewarded.

While I am not discounting the contributions of the thousands of hard-working public school teachers across the nation, who selflessly work to educate and uplift the children under their care, the dismal working conditions associated with the job means that the majority of Indian students become victims of a system that is out of their control.

While government spending has seen a gradual decline, the cost of education in India is rising year upon year.

As of a 2019 survey conducted by CARE, average household spending for professional courses can set a

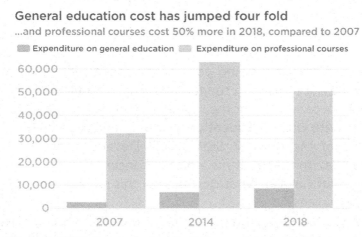

Figure 10-6: Rise in cost of education.[15] Source: Business Insider

person back Rs 64,763 in urban areas, a number that is 2x that of rural areas, where the price is Rs 32,137. The report further states that expenditures can rise as high as Rs 72,000 per student.[16]

Of this, 20 per cent of education-associated expenditures go towards the purchase of books, stationery and school uniforms.

Adding to the problem, the further a student studies, the higher the cost of education climbs. These escalating costs and a general lack of accessible and affordable options are likely the reason 29 per cent of girls and boys drop out of school before completing their elementary educational requirements, with the demographic worst affected being children from marginalised communities.

As per the 2011 census, Indian males showed higher rates of literacy than their female counterparts, scoring 82.14 per cent compared to 65.46 per cent for females.[17] [18] While the Indian government's many initiatives to educate the girl child do indeed represent a positive trend, this figure is indicative of the marginalisation of girl children in rural areas. These young girls, not being considered capable of earning money or furthering the family name, are pulled out of school to work at home, or to be married off.

With Quality Education and Gender Equality being two of the UN's 17 Sustainable Development Goals (SDGs), this disparity is a global concern.

'But Aakash, what does any of this have to do with my choices?'

So far, I have painted a picture of the gaps in our education infrastructure. Now, let's take a step back

stream	percentage		
	male	female	person

Percentage distribution of students pursuing technical/professional course by type of course/ stream

all-India

stream	male	female	person
rural			
medicine	3.8	9.2	5.5
engineering	30.1	20.2	27.0
agriculture	4.1	3.0	3.8
law	1.2	0.7	1.1
management	2.7	6.9	4.0
education (B.Ed, M.Ed, etc)	7.2	21.2	11.6
CA and similar courses*	0.6	1.7	0.9
IT/computer course	9.5	10.4	9.8
courses from ITI/recognised vocational institutes	30.4	10.3	24.0
others	10.3	16.4	12.2
all	100.0	100.0	100.0
urban			
medicine	5.9	16.9	10.0
engineering	51.6	33.6	44.9
agriculture	1.6	1.6	1.6
law	2.1	2.6	2.3
management	6.4	7.8	6.9
education (B.Ed, M.Ed, etc)	3.6	10.1	6.0
CA and similar courses*	3.1	2.6	2.9
IT/computer courses	8.9	9.7	9.2
courses from ITI/recognised vocational institutes	11.1	3.7	8.3
others	5.8	11.3	7.9
all	100.0	100.0	100.0
rural+urban			
medicine	4.9	13.8	8.0
engineering	41.6	28.2	37.0
agriculture	2.7	2.1	2.5
law	1.7	1.8	1.8
management	4.7	7.4	5.6
education (B.Ed, M.Ed, etc)	5.3	14.6	8.5
CA and similar courses*	1.9	2.2	2.0
IT/computer courses	9.2	10.0	9.5
courses from ITI/recognised vocational institutes	20.0	6.4	15.3
others	7.9	13.4	9.8
all	100.0	100.0	100.0
*like Company Secretary (CS), Cost and Works Accountants (CWA), etc.			

Table 10-2: Percentage distribution of students pursuing technical/ professional type of courses.[19] Source: Government of India

and discuss the choice of curriculum amongst students in India.

While many students reading here would proclaim enthusiastically that they do not get a say in their educational and career decisions, and while this may indeed be the case, we can only choose between the options provided to us, which presents a problem in and of itself.

As the age-old joke indicates, most students in India are forced to pursue engineering, science, medicine or law—the most statistically lucrative educational choices. In even the most urbanised parts of the country, the decision to pursue creative fields, such as art, music or theatre, is considered a sign of a child's poor intellect.

At the post-graduate level, 6.55 lakh students are currently (2022) enrolled in science courses, which are further divided into eighteen sub-streams. The highest enrolment of these sub-streams is in mathematics, with a total of 1,43,116 students. Chemistry follows with 1,36,807 students, Physics with 76,382 students and zoology with 59,045 students.[20]

The reason I mention this is because there is another scientific field that has been largely ignored in India: environmental science.

Education for the Future: Environmental Education

While environmental education (EE) has been made a compulsory subject in educational institutions following multiple rulings by the Supreme Court of India, the subject received little interest from youngsters in its early days. Not much has changed since then.

This comes as little surprise as curious young minds and educators alike tend to view such compulsory inclusions as less important compared to other subjects; they lack the passion and enthusiasm that could inspire greater interest in the subject.

However, EE, in my opinion, is perhaps the most important subject for the coming generations, who will bear the brunt of our environmental negligence.

Alison Barrett MBE, British Council, director, India, says:

'Teaching climate change should not be left to scientists or science teachers, but it should be solutions-oriented and practical. Through an active and experiential approach, education can help break the cycle of cultural practices. Communication, language, influencing, and media skills are essential for our future success. Whether that's scientists or journalists communicating complex research to the wider public or teachers communicating to their students or artists communicating their ideas through multi-media, or activists advocating for change or negotiating deals. This means making global and indigenous knowledge accessible to all, and in diverse, engaging, and age-appropriate formats—not just scientific texts but stories, poems, or songs. Non-verbal approaches to knowledge building may have the greatest impact.'

EE has been widely acknowledged to be the most effective tool for environmental conservation, promoting not just general knowledge of our environment, but also instilling the value of green social behaviours among the youth.[21]

While we have had some minor success in implementing EE curriculums in schools, there is a multitude of challenges to its adoption in India.

Let's take a look at some of these challenges now.

Lack of Comprehensive Content

As the situation currently stands, compulsory EE curriculums are limited in their comprehensiveness and ability to effectively inform.

Going beyond the theory of the matter, students must be allowed to practically experience the value of this knowledge, through regular interactions with the natural world. Granted, it is not uncommon to see school children engaging in beach clean-ups or plantation activities on World Environment Day; but this is more a singular event with a highly performative nature that is viewed more as a recreational outing than an educational experience by teachers and students.

If we are not made aware of how interconnected our lives are with the planet, we will never be able to change our ways to preserve the longevity of both.

For example, when I was younger, I had a keen interest in science, and would often ask my teachers questions about the way the world works that couldn't quite be covered by my textbooks.

Every single time, I would return empty-handed as my curiosity was met with indifference. 'Study what's in the book, no extra questions,' my teachers would say in a reprimanding tone.

To them, my exam score mattered more than my curiosity.

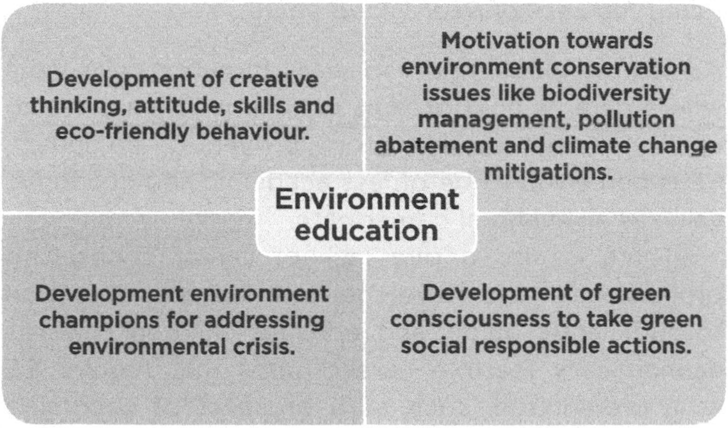

Development of creative thinking, attitude, skills and eco-friendly behaviour.

Motivation towards environment conservation issues like biodiversity management, pollution abatement and climate change mitigations.

Environment education

Development environment champions for addressing environmental crisis.

Development of green consciousness to take green social responsible actions.

Figure 10-7: The effectiveness of environment education. Source: Government of India[22]

Time and Importance

Environmental education is widely perceived as a subject that is less important than others and is typically an elective chosen as an addition or substitution that requires less effort from students.

On average, compulsory EE classes occur once or twice a week, while subjects like mathematics have classes every single day. Moreover, higher electives of the same subject are also offered. What this indicates is that our institutions do not prioritize EE as a part of their annual timetables.

On the other hand, students in high school and above are typically overburdened by the curriculum of their other courses and extra-curricular activities and do not have the time or patience to think deeply about our natural world.

Perceived Lack of Real-World Value

Across the globe, and especially in nations like India where parents hold dreams of their children growing up to be celebrated doctors, lawyers or engineers, EE is perceived to be of less academic and economic value as an eventual career path.

There is also a dire lack of career counselling professionals who have the knowledge or intention to sufficiently guide and educate children towards economically feasible career paths that require EE as a prerequisite, such as a prospect of becoming an environmental advisor at large, corporate firms or that of becoming a specialist at any number of internationally recognised institutions such as United Nations Development Programme, World Wildlife Fund and more.

Eligibility and Lack of Training Amongst Educators

Indian teacher training courses, such as B.Ed and D.El. Ed, hardly mention EE as a point of discussion. This means that most educators are ill-informed about the subject they are made to teach as a mandatory inclusion in their pupils' curriculums.

Additionally, the eligibility criteria of EE teachers in India are a murky affair. Any individual with a postgraduate degree in science is considered eligible for the job—even though this is not necessarily indicative of their knowledge of the subject.

Lack of Societal Interest and Institutional Support

While education is a matter of national significance, legislators, teachers and students have additional societal responsibilities to fulfil, meaning that even the basic knowledge of the complex fabric of our environment suffers.

In much the same way as we are taught that crime is wrong and that charity is admirable, every member of society has a responsibility to educate the coming generations on the importance of environmental conservation and sustainability.

According to Sanjay Seth, senior director, sustainable infrastructure programme, The Energy and Resources Institute (TERI) and vice-president and CEO of Green Rating for Integrated Habitat Assessment (GRIHA) Council:

'Climate Change can't be fixed overnight. Whilst the global community is working tirelessly towards finding the best remedial solutions and implementing them, environmental education shall prove to be the cornerstone in the climate action that will shape the thought processes of our future decision-makers. Sustainable Living should not be plug-in, but inherent. environmental education at an early age can help breed curiosity, create awareness regarding environmental issues and nurture a passion for mother nature. With the vision to create a net zero economy by 2070, the green business sector is also immensely promising for the potential and the ripened minds opting for courses in environmental and ecological science. The field has diverse job opportunities that range from environmental scientists to filmmakers, photographers, engineers, and so on.'

While the effective adoption of EE is indeed a challenge, our species must take it as a serious challenge to ensure that future generations are afforded a chance at a good quality of life and can partake in their natural heritage in the same way we have.

So, why wait for educational institutions to change? Each of us has a choice to educate and inspire the next generations of learners, teachers and creators. With the internet making useful research available to us at the click of a button, there is no lack of opportunities for us to educate ourselves and others on the value of sustainability.

But what will it take to make a change on a societal level?

Human beings are free-thinking, independent beings, but as we have learned, we lack direction and purpose. When caught in a raging river, the wise man moves with the flow, but if the river is barrelling forward towards a waterfall, we must learn to brave the tide and change our direction.

However, going with the flow is the path of least resistance and unless we are properly informed of the impending danger, few will choose to expend the effort necessary. If we believe we are safe, and that things are as they should be, we will never see the drop coming.

But to be properly informed, we need two things: someone with a bird's eye perspective of the situation that can see what lies ahead before we do, and for that person to be a trustworthy guide, who cares deeply about the people caught in the currents.

Spin Doctors and Red Tape: The Problem with Politics

Democracy, the system of governing that reflects the will of the people, has been widely accepted as the most ideal governing structure for global societies. Originating in Athens in the fifth century BC, today, democracy today rules the world — today over half of the countries across the world are being governed by democratic structures.[23]

While democratic governance sounds ideal in theory, in practice, global populations have expressed their dissatisfaction over the functioning of these governmental organisations.[24]

If democracy is a form of government that is truly for the people, by the people and of the people, then what causes this disparity between what the people want and what the government delivers?

While I am no political expert, I believe my answer is a satisfactory one for any country that is suffering under the guise of a fair and equal democratic system. If you were to ask me what the biggest problem with politics is, I would tell you it is the same, as is for other capitalistic structures in the world: greed.

India is known for its large-scale political campaigns, which have only become extravagant in recent times.

Divided views around the world about how key aspects of democracy are working

% who say they are ___ with the way democracy is working in their country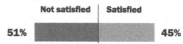

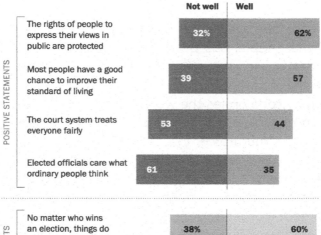

Figure 10-8: World view on democracy.[25] Source: Pew Research Centre

From congestion on roads blocking the passage of ambulances so that some minister can travel without hassle to staged coups and communal uprisings that conveniently coincide with election season, we Indians are no strangers to the games these elite masters play in the name of democracy.

Whether it is political power, money or social standing, it has become exceedingly clear to the civil population that those in power do not always have our best interests in mind. From ex-president Donald Trump's infamous march on the Capitol building on 6 January 2021 after having lost the election to the current US President Joe Biden, to the rise of Hitler, who used existing loopholes in the democratic system to gain dictatorial control over Germany, leading to perhaps the most devastating war ever witnessed in modern memory—there is no lack of examples of democracy falling into the hands of power-hungry megalomaniacs.

But how can we ensure that the wrong people do not come to power? The obvious answer is to vote and to vote wisely. We must invest in learning about these figures, what they stand for and what they promise, and choose our votes wisely. We must raise our voices against those who promise sustainable solutions and then pivot for greed. We must invest in our governments and expect the same from them.

However, the global call to arms for us to change our voting decisions is just another form of a capitalistic structure pushing the burden of reparation onto the consumer.

How, you ask?

According to a survey[26] of politicians in three Indian states, more than 60 per cent state and national officials reported receiving political party support compared to fewer than 10 per cent of respondents at the district level and lower. Adding to this, significant contributions are received at the state and national level from a variety of sources with some being less legally sound than others.

This information indicates that political parties at the state and national levels hold great sway over representatives in other areas and further along the food chain. While those who align with the party's stances on any number of subjects are paid handsomely for their service, individuals running for a term in office at the regional level must rely on personal funding for their campaigns.

This means that no matter how good you are, or how much your vision for our nation could change for the better of our people, your voice might be drowned out by that of another who has less vision, but more capital on hand.

In real-world terms, this means that ground-level changemakers, who have seen the problems plaguing our nation and our people with their eyes, and who have done the work to find the right legislative solutions, may never even make it to the ballots due to a lack of political funding.

However, democratic structures also advocate for self-determination and free speech, meaning that change is possible, but only if the people advocate for it. From the widespread Sri Lankan unrest in 2022 against the powers that be, to the Iranian civil protests for equal rights for all genders, people across the globe are raising their voices to be heard by a system that seems to be built to ignore them.

But a system powered by the people cannot ignore them for long; we must hold our governing bodies responsible for their actions.

As employees of the nation, our governments function on taxpayer money, meaning that they work for *you*. The sooner we, the general population, can understand and absorb this concept, the sooner we can enact real change in our nations and the world.

But what about us, who do we truly work for?

Green Money—Sustainable Careers

As we have already discussed in the education section of this chapter, Indian children are expected to make careers in the medicine, law or engineering. Climate and environmental studies are considered throwaway courses with a lack of opportunity for the future.

However, this is simply not the case in the 2020s. With the rise of companies advocating for better environmental, social, and governance (ESG) standards for their operations and with the rising interest in ecological conservation worldwide, more and more people are choosing to make a difference with their careers.

Shradha Sharma, founder and CEO of YourStory, says:

'Our market is primed and ready for environmental and ecological jobs. We have one of the youngest working-age populations globally and what's increasingly clear is that the big job creators of yesterday and today will not be exactly the same tomorrow. As environment and ecology become conversations, front, and center, jobs emerging from these and skills people need to build to get ready for these jobs, become more critical. Not just that, our recent New Education Policy has created more flexible pathways of higher education studies which in itself can also be a great fillip for a journey towards newer and more non-traditional jobs. Investors choosing to invest in this area, need to build their strong domain expertise and commit to the long term.'

All around the world, consumers are demanding more sustainable products that are not just ecologically sound in their applications, but also in their manufacturing, as shown by this graph of companies reporting on their sustainability from the year 1993 onwards.

Nikhil Taneja, CEO and co-founder, Yuvaa, says:

'We are still not doing enough as a world and society to address climate change, to alleviate the concerns of Gen Alpha (and even a lot of Gen Zs), and we are sending a message that we don't really care about what happens to the young people who will inherit the earth, and it's ultimately all in their hands, which is a massive responsibility and burden for them in their growing years. The measures young people can take are the ones they are already taking: making themselves heard, standing up for the climate cause, being loud, being impatient, and being aggressive

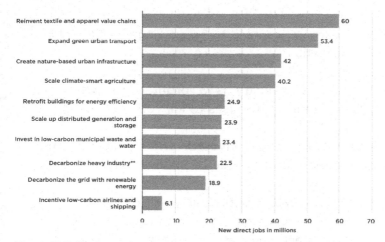

Figure 10-9: Green employment opportunities between emerging markets 2020-2030.[27] Source: Statista

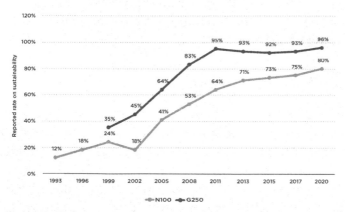

Figure 10-10: Companies that report on sustainability from 1993 to 2020.[28] Source: Statista

about change. There's no time to waste and young people are aware of that. We need to understand that it's the world of the young, we are just living in it—and it's important that we do everything in our power to make it more habitable for them. That's the legacy we need to be working on as the older generation.'

Today, sustainability is a profitable industry, while unsustainability promises to kill your business.

Ritesh Malik, angel investor and founder of Innov8, and founder and trustee of Plaksha, says:

'In this climate, sustainability and conservation are not just important, but vital to the survival of our species and countless others. The perception that climate change is someone else's problem holds little sway in today's data-driven world, and investors are on the lookout for the next green unicorn to take the world by storm.'

What Are We Missing?

Belief can change the world. The choices we make every single day define who we are, who we will become and what we will leave behind for the next generation to learn from.

People always say it is impolite to talk about religion and politics with people you do not know. I, for one, disagree.

If we cannot talk about our beliefs and choices, how are we ever meant to relate to each other, to understand what divides us, so that we can look for solutions to unite?

For example, modern media would have us believe that religious belief is a source of hate and subjugation. But, at their root, all religions advocate for peace and conservation.

Let's take the three faiths we discussed at length at the beginning of this chapter and take a closer look at their relationship with nature.

Hinduism

A religion that is founded in the pagan belief system of worshipping the natural elements, Hindu sacred texts, such as the Vedas to the great epics all make specific mention of natural divinity. Thus there are gods of wind, thunder and sunlight to the spirits of rivers, mountains, trees and animals.

At its core, Hinduism advocates to protect and revere the natural world. However, belief is only half the battle fought. We must choose every day to honour our beliefs and make decisions that benefit the majority over the minority.

'According to the different modes of material nature — the mode of goodness, the mode of passion and the mode of darkness — there are different living creatures, who are known as demigods, human beings and hellish living entities. O King, even a particular mode of nature, being mixed with the other two, is divided into three, and thus each kind of living creature is influenced by the other modes and acquires its habits also,' says the Bhagavata Purana (2.10.41)

Christianity

Christian faith draws its tenets from the Holy Bible, a scripture that contains over a hundred verses regarding the importance to protect our environment and our sense of morality.

A religion that takes its roots in forgiveness and atonement of one's sins, the Christian faith, in its truest form, understands the gravity of our choices, and what it would take to rectify the damage done.

'Do not pollute the land where you are. Bloodshed pollutes the land, and atonement cannot be made for the land on which blood has been shed, except by the blood of the one who shed it,' says the Bible (verse 35:33).

Islam

A religion that views humanity as the protectors of the creations of God, the Islamic faith not only advocates for sustainability and conservation, but also humility, respect, and a waste-not-want-not way of life.

'It is Allah who made for you the earth a place of settlement and the sky a ceiling and formed you and perfected your forms and provided you with good things. That is Allah, your Lord; then blessed is Allah, Lord of the worlds,' says Quran (40:64).

Islam teaches us not to be harbour pride or greed and to respect the awesome power of the natural world.

'Do not strut arrogantly on the earth. You will never split the earth apart nor will you ever rival the mountains' stature,' says the Quran (17: 37).

And it's not just these religions that say it.

From the Jain scholars Mahavira's mandate, 'Do not injure, abuse, oppress, enslave, insult, torment, torture, or kill any creature or living being', to the words of the Genesis 1:29 of Judaism that go, 'God said: "Behold, I have given you every herb yielding seed, which is upon the face of all the earth, and every tree, in which is the fruit of a tree yielding seed--to you, it shall be for food"', all religions always advocate for more than modern capitalistic tradition would have us believe.

Much like religious belief, capitalism has made quick work of societal structures that were built to maintain peace, unity and preservation of natural resources, perverting them until we are left with nothing but a deteriorating natural and manmade world, with a hefty bill attached that must be paid out of pocket.

But why do we pay for our beliefs? What is the true cost of our decisions and who reaps the benefits?

Our religious observances, each one a celebration of light, love and life, serve to make so many of us engage in behaviours that are harmful to the environment, which, by their own teachings are against the true meaning of their founding tenets. Each of these religions has something to teach us about living in harmony with nature, if only we take the initiative to learn and listen.

However, we must also understand that the world has changed drastically since these institutions were first founded. The world population increased from one billion in 1800 to around eight billion today.[29] This means more mouths to feed, fewer resources left to consume and more emissions per person than have ever been recorded across human history.

In fact, one of the lesser-known but most effective ways of reducing your carbon footprint is to have fewer children.[30]

Writing this book, and this chapter, in particular, taught me two very important lessons—the first being that there is a significant gap between what we think we know and what information is readily available for public consumption, and the second, is that our generation is in dire need of good people doing good work, and braving the odds to help educate and inspire change among their people.

Humanity is at a crossroads and which way we choose to go will reflect our beliefs.

So, how can we make better choices?

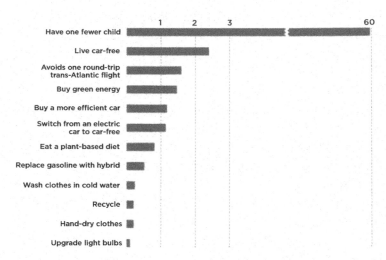

Figure 10-11: Emissions savings (tonnes of carbon dioxide equivalents).[31]
Source: Science

How to Change Your Mind

Ask yourself this: what do you care about?

While some of you will immediately think of your family, your bank account, your education, or even just what's for dinner tonight, what we all tend to forget when we prioritize our beliefs and choices is this—everything you care about, have ever cared about, or will care about over the course of your life, is at the mercy of the natural world.

Nithin Kamath, co-founder and CEO, Zerodha, says:

> '"ESG" as a trend has only caught on in the last five-odd years in developed markets, and it isn't part of the conversation in developing markets like India yet. But this will change over the years. Arguably, ESG investing has no real-world impact. The stock markets are the wrong place to make a difference, but there's a certain "warm-glow" effect of knowing that your investments align with your beliefs, and that will catch on over the years. On a broader level, India needs more companies thinking locally, sourcing locally, and producing locally, ensuring that jobs are created locally and that the entire supply chain is green and aligned with the local ecologies. If we see more companies and brands thinking along these lines, I believe their customers will follow suit.'

This is because we are not mere spectators in the grand theatre of life but are a *part* of it. We have to stop thinking of the environment as something that is outside of us because it *is* us. It is our past, our future, and the only present we will ever see. By killing it we are killing ourselves.

Our beliefs guide us, and if we do not believe in the future, there will be no future left for anyone to believe in.

So, I ask you again: what is it that you truly care about?

Resources

Here is an inventory of supplementary media material, people who matter and shops in line with the chapter's theme.

Books

1. *Braiding Sweetgrass: Indigenous wisdom, scientific knowledge and the teaching of plants*
    Author: Robin Wall Kimmer
    Published by: Penguin
    https://www.amazon.in/Braiding-Sweetgrass-Indigenous-Scientific-Knowledge/dp/1571313567

2. *The Nature Fix- Why Nature Makes us Happier, Healthier, and More Creative*
    Author: Florence Williams
    Published by: W.W. Norton & Company
    https://www.amazon.in/Nature-Fix-Happier-Healthier-Creative/dp/0393242714

3. *10 Indian Champions: Who are Fighting to Save the World*
    Author: Bijal Vachharajani
    Published by: Penguin Random House
    https://penguin.co.in/book/10-indian-champions/

4. *Race for Tomorrow: Survival, Innovation, and Profit on the Front Lines of the Climate Crisis.*
    Author: Simon Mundy
    Published by: William Collins
    https://www.amazon.in/Race-Tomorrow-Scramble-Survival-Changing/dp/000839430X

5. *This Changes Everything: Capitalism vs. The Climate*
   Author: Naomi Klein
   Published by: Penguin
   https://www.amazon.in/This-Changes-Everything-Capitalism-Climate/dp/0241956188

Documentaries

1. *Lowland Kids*
   Directed by: Sandra Winther
   Available on Youtube
   https://www.youtube.com/watch?v=tHV0F3-72Eo

2. *Before the Flood*
   Directed by: Fisher Stevens
   Available on Natgeo, Disney+Hotstar
   https://www.beforetheflood.com/

3. *Don't Look Up*
   Directed by: Adam Mckay
   Available on Netflix
   https://www.netflix.com/in/title/81252357

4. *Extract.Destroy.Repeat*
   Directed by: Waterbear Team
   Available to watch for free on Waterbear
   https://www.waterbear.com/watch/short/636cbcd3483684a8103b5f91

5. *So Hot Right Now*
   Directed by: Waterbear Team
   Available to watch for free on Waterbear

https://www.waterbear.com/watch/6048adee02476cc0749dceb2

Ted Talks

1. It's Impossible to Have Healthy People on a Sick Planet
   By: Shweta Narayan
   https://www.youtube.com/watch?v=ZOsHRaEhiAQ

2. The Disarming Case to Act Right Now on Climate Change
   By: Greta Thunberg
   https://www.youtube.com/watch?v=H2QxFM9yOtY&t=81s

3. The Fastest Way to Slow Climate Change Now
   By: Ilissa Ocko
   https://www.youtube.com/watch?v=tIWuP7wESZw&t=136s

4. The Most Important Thing You Can Do to Fight Climate Change: Talk About IT.
   By: Katharine Hayhoe
   https://www.youtube.com/watch?v=-BvcToPZCLI

5. What Nature Can Teach us About Sustainable Business
   By: Erin Meezan
   https://www.youtube.com/watch?v=TxwGZppT2WA&t=14s

People

1. Saif | @thesustainabilityguy
   https://www.instagram.com/thesustainabilityguy

2. Nayana | @nayana_premnath
   https://www.instagram.com/nayana_premnath

3. Pankti Pandey | @Zerowasteadda
   https://www.instagram.com/zerowasteadda

Experiences and Services

1. Greenmyna
   Eco-consultants for weddings and events
   https://www.greenmyna.com/

2. Skrap Zerowaste
   Zero waste and sustainability services for businesses
   https://www.skrap.in/

3. Centre for Environment Education
   Courses for everyone to learn about climate change, sustainability, and the environment
   https://www.ceeindia.org/

4. Sadhna Forest
   Community working for ecological revival and sustainable living
   Anyone can Volunteer/ stay to experience a sustainable way of life
   https://sadhanaforest.org/

5. Auroville

   The city of Dawn offers a lot of workshops on sustainable living, green practices, natural healing, and many more that take us closer to mother nature and urge us to protect her.

   https://auroville.org/

Acknowledgements

Life on the road for close to a decade—and there are stories that cannot even begin to describe the experiences, dreams and vision for a sustainable future that has formed in this time. Yet, I have tried to etch my learnings in some pages so that I open to people a world beyond concrete jungles—a world that needs saving so that it can love you back with the strength that it was always meant to. Realizing this vision was not possible without a team that just fell in place. This book is theirs as much as it is mine and I would like to thank them very sincerely:

- Premanka Goswami: Associate publisher of Penguin Random House India
- Yash Daiv: Associate copy editor of Penguin Random House India
- Anchal Goil: Wordsmith
- Sruthi Subbanna: Researcher

- Tanishka Sharma: Project manager
- Sugandha Kharya and Amit Meena: Cover designers
- Saurabh Podey: Digital consultant
- Nividha Vij: Graphic designer

*

I would like to thank my champions for their unwavering support—Archana Jain and Pareina Thapar for the unparalleled love and encouragement they have given to me on this book and otherwise. I tip my hat to you both!

*

I would like to thank my family, who sheltered me, feed me, loved me and did everything they thought would help me become a better person.

My gratitude to: Manorama Shukla, Sapna Shukla, Sarita Shukla, Sanjay Shukla.

*

I would like to thank my friends, who have trusted me and have been by my side cheering me from the sidelines. They have picked up my phone calls at random hours to listen to me and my ideas and my confusion, too. They have invested their time and energy and have done everything in their power to support my dreams and shower me with their love and friendship. Thank you:

Sonal Ved, Sidharth Rajhans, Shreya Punj, Shefali Deshwali, Tarun Gupta, Sapna Bhatia, Ratika Yashwante, Sidhant Sidana, Shashank Vaishnav, Parveen Singhal, Vinay Singhal, Shradha Sharma, Kanika Joshi, Esha Tiwari, Yash Kotak, Manish Pandey, Arpita Nath, Barry Rodgers, Anushka Sani, Aakanksha Gupta, Vasudha Rai, Tanaya Narendra, Sohni Tabassum, Gunjan Jain, Shruti Anand, Parth Bhalerao.

A Big Thank You To Industry Experts

I would like to thank industry experts who have given their inputs for the book. These influential people, who employ millions of youngsters across the country, vouch for sustainability. There is no bigger honour that they stood with me as I tried to relay my message to the world. Their inputs about fighting climate change and advocacy of sustainability have added immense value to the book, making it unique. Their thought processes aligned with the book and optimism and will be always grateful for that.

- Piyush Pandey, chief creative officer worldwide and executive chairman India of Ogilvy & Mather; Padma Shree Awardee
- Dia Mirza, United Nations Environment Programme Goodwill Ambassador and actress

- Keegan Kuhn, co-director of *Cowspiracy, What the Health*
- Sonal Ved, author and digital editor, *Cosmopolitan*, *Harper Bazaar* and *Bridal India*
- Sohil Wazir, chief commercial officer, Blue Tribe Foods
- Manoj Kumar, co-founder, Araku Coffee
- Dr Kiran Ahuja, manager of vegan projects, PETA India
- Yash Kotak, co-founder of Bombay Hemp and Co.
- Taran Chhabra, founder, Neeman's Shoes
- Rahul Mishra, luxury fashion designer
- Apurva Kothari, founder of No Nasties
- Mehernaaz Dhondy, editor-in-chief, *Grazia*
- Santosh Iyer, MD and CEO, Mercedes-Benz India Pvt Ltd
- Sandeep Singh, managing director of Alkem Laboratories Ltd
- Anmol Jaggi, co-founder of Blusmart
- Ramesh Somani, editor-in-chief, *TopGear*
- Mahesh Babu, CEO, Switch Mobility
- Shivya Nath, author, founder, climate conscious travel
- Ishita Khanna, co-founder and director, Spiti Ecosphere
- Malika Virdi, founder and CEO, Himalayan Ark
- Aindrila Mitra Rajawat, editor-in-chief, *Travel & Leisure*
- Dr Tanaya Narendra aka Dr Cuterus, doctor, author
- Mandeep Manocha, founder and CEO, Cashify

- Nikhil Arora, thought leader and CEO
- Ashish Singhal, co-founder and CEO, CoinSwitch
- Tarun Jami, founder, GreenJams
- Aakarsh Nayyar, founder, Aliste Technologies
- Vani Murthy, urban farmer, founding member, SWMRT Bengaluru
- Chaitsi Ahuja, founder and CEO, Brown Living India
- Jessica Jayne, founder, Pahadi Local
- Samath Bedi, executive director, Forest Essentials
- Vasudha Rai, author
- Shriti Malhotra, CEO of The Body Shop,
- Prianka Jhaveri, founder, The Mend Packaging
- Gunjan Jindal Poddar, founder, Amala Earth
- Nandan Bhat, founder and director, EcoKaari
- Pradeep Sangwan, founder, Healing Himalayas
- Sahar Mansoor, co-founder, Bare Necessities
- Ashwin Malwade, founder, Green Myna
- Ritesh Malik, angel investor, founder of Innov8, founder and trustee of Plaksha
- Sanjay Seth, senior director of Sustainable Infrastructure Programme, The Energy and Resources Institute (TERI) and vice-president and CEO of Green Rating for Integrated Habitat Assessment (GRIHA) Council
- Shradha Sharma, founder and CEO, YourStory
- Nikhil Taneja, co-founder and CEO, Yuvaa
- Nikhil Kamath, co-founder and CEO, Zerodha
- Alison Barrett MBE, British Council, director, India

Notes

Chapter 1

1   'The Scale and Drivers of Carbon Footprints in Households, Cities and Regions Across India', Research Institute of Humanity and Nature, accessed 6 March 2023, https://www.chikyu.ac.jp/rihn_e/

2   'Addressing Climate Change Concerns in Practice', American Psychological Association, accessed 6 March 2023, https://www.apa.org/monitor/2021/03/ce-climate-change

3   'Religion and Food', Pew Research Centre, https://www.pewresearch.org/religion/2021/06/29/religion-and-food/#:~:text=Altogether%2C%20about%20eight%2Din%2D,%25)%20are%20the%20least%20likely, accessed 6 March 2023

4   'What India Can Teach the World About Sustainability', World Economic Forum, https://www.weforum.org/agenda/2017/10/what-india-can-teach-the-world-about-sustainability/, accessed on 6 March 2023

5   See note 4

6   'The CAT Thermometre Explained', Climate Action Tracker,https://climateactiontracker.org/global/cat-thermometer/, accessed on 6 March 2023

7 'No New Coal by 2021: The Collapse Of The Global Coal Pipeline', Global Energy Monitor, https://globalenergymonitor.org/wp-content/uploads/2021/09/No-New-Coal-by-2021-the-collapse-of-the-global-pipeline.pdf, accessed on 6 March 2023

8 'Lazard's Levelized Cost of Energy Analysis—Version 14.0', Lazard, https://www.lazard.com/media/451419/lazards-levelized-cost-of-energy-version-140.pdf, accessed 6 March 2023

9 'Renewable Energy', Our World in Data, https://ourworldindata.org/renewable-energy, accessed in 6 March 2023

Chapter 2

1 'New FAO analysis reveals carbon footprint of agri-food supply chain', United Nations, https://news.un.org/en/story/2021/11/1105172, accessed on 28 October 2022

2 'Food and Grocery Retail Market Size', Grand View Research, https://www.grandviewresearch.com/industry-analysis/food-grocery-retail-market, visited on 28 October 2022

3 'Food Security and Nutrition Around the World in 2020', Food and Agriculture Organization of the United Nations, visited on 28 October 2022

4 'Seeking end to loss and waste of food along production chain', Food and Agriculture Organization of the United Nations, visited on 28 October 2022

5 'UNEP Food Waste Index Report 2021', United Nation Environment Programme, visited on 28 October 2022

6 'The State of Food Security and Nutrition in the World, 2017', Food and Agriculture Organization, https://www.fao.org/agrifood-economics/publications/detail/en/c/1036736/#:~:text=This%20year's%20The%20State%20of,of%20the%20above%2Dmentioned%20factors.

7 Publisher: George Allen and Unwin (original), Puffin Books (1995–2006), Scholastic (current)

8   'Environmental Impacts of Chocolate Production and Consumption in the UK', Food Research International, Science Direct, https://www.sciencedirect.com/science/article/abs/pii/S0963996918301273#:~:text=Chocolate%20consumption%20in%20the%20UK,whole%20food%20and%20drink%20sector, accessed on 28 October 2022

9   'Is Chocolate Bad for the Environment?', Envirotech Online, https://www.envirotech-online.com/news/business-news/44/breaking-news/is-chocolate-bad-for-the-environment/45837, accessed on 28 October 2022

10 'Impact of Chocolate on our Climate', News Medical, https://www.news-medical.net/health/Impact-of-Chocolate-on-our-Climate.aspx, accessed on 28 October 2022

11 'Chocolate's Dark Secret', Mighty Earth,https://www.mightyearth.org/wp-content/uploads/2017/09/chocolates_dark_secret_english_web.pdf, accessed on 28 October 2022

12 'Impact of Chocolate on our Climate', News Medical, https://www.news-medical.net/health/Impact-of-Chocolate-on-our-Climate.aspx, accessed on 28 October 2022

13 'Chocolate Makers Have Sweet-Loving Indians in their Sights', Bloomberg, https://www.bloomberg.com/news/articles/2021-03-31/chocolate-makers-aim-to-win-over-india-s-sweet-loving-consumers, accessed 28 October 2022

14 'Cocoa industry is pinning growth hopes on India's sweet tooth', International Cocoa Organization (ICCO), https://www.livemint.com/news/world/cocoa-industry-is-pinning-growth-hopes-on-india-s-sweet-tooth-11614603881571.html, accessed on 28 October 2022

15 'How Much Rainforest Is in that Chocolate Bar?', World Resource Institute, https://www.wri.org/insights/how-

much-rainforest-chocolate-bar, accessed on 28 October 2022

16　'Greenhouse Gas Emissions from the Dairy Sector', Food and Agriculture Organization of the United Nations—Animal Production and Health Division, https://www.fao.org/3/k7930e/k7930e00.pdf, accessed on 28 October 2022

17　*Cowspiracy* directed by Keegan Kuhn and Kip Andersen, https://www.cowspiracy.com/, accessed 28 October 2022

18　'Methane: A crucial opportunity in the climate fight', Environmental Defense Fund, https://www.edf.org/climate/methane-crucial-opportunity-climate-fight, accessed 28 October 2022

19　'CO2 Equivalents', Climate Change Connection, https://climatechangeconnection.org/emissions/co2-equivalents, accessed on 28 October 2022

20　'Questions and answers about palm oil', Rainforest Rescue, https://www.rainforest-rescue.org/topics/palm-oil/questions-and-answers, accessed on 28 October 2022

21　'Food: Greenhouse Gas Emissions Across the Supply Chain', Our World in Data, https://ourworldindata.org/grapher/food-emissions-supply-chain, accessed on 28 October 2022

22　'Global Food System Emissions Could Preclude Achieving the 1.5° and 2°C Climate Change Targets', Science, https://www.science.org/doi/10.1126/science.aba7357, accessed on 28 October 2022

23　'Livestock and Climate Change', ResearchGate, https://www.researchgate.net/publication/285678846_Livestock_and_climate_change, accessed on 28 October 2022

24　'Global food system emissions could preclude achieving the 1.5° and 2°C climate change targets', National Library of Medicine, National Center for Biotechnology Information, visited on 28 October 2022, https://pubmed.ncbi.nlm.nih.gov/33154139

25 'Environmental Footprints of Dairy and Plant-based Milks', Our World in Data, https://ourworldindata.org/grapher/environmental-footprint-milks, accessed on 28 October 2022

26 'Water Use in Livestock Production Systems and Supply Chains', Food and Agriculture Organization of the United Nations, https://www.fao.org/3/ca5685en/ca5685en.pdf, accessed on 28 October 2022

27 *Milked* documentary, directed by Amy Taylor, https://milked.film/, accessed on 28 October 2022

28 'Breeds of cattle—Animal Husbandry', TNAU Agritech Portal, https://agritech.tnau.ac.in/ta/animal_husbandry/animhus_cattle%20_breed.html, accessed on 28 October 2022

29 'Climate Crisis: Way Forward for Dairy Giants in India', Down to Earth, https://www.downtoearth.org.in/blog/climate-change/climate-crisis-way-forward-for-dairy-giants-in-india-78226, accessed on 28 October 2022

30 'High-Growth Segments of Indian Food and Beverage Industry—World Food India 2017', Ministry of Food Processing Industries Government of India, http://www.worldfoodindia.gov.in/2017/images/high-growth.pdf, accessed on 28 October 2022

31 'Opportunities in Meat and Poultry Sector in India', Ministry of Food Processing Industries, https://www.mofpi.gov.in/sites/default/files/OpportunityinMeat%26PoultrysectorinIndia.pdf, accessed on 28 October 2022

32 'Global Beef Update: Exporters', Drovers, https://www.drovers.com/news/beef-production/global-beef-update-exporters, accessed on 28 October 2022

33 'How Is India's Meat Industry Impacting the Environment?', India Bioscience, https://indiabioscience.org/columns/indian-scenario/meating-the-needs-of-the-future, accessed on 28 October 2022

34 'Trends and Projected Estimates of GHG Emissions from Indian Livestock in Comparisons with GHG

Emissions from World and Developing Countries', Asian Australasian Journal of Animal Sciences—ResearchGate, https://www.researchgate.net/publication/264128386_Trends_and_Projected_Estimates_of_GHG_Emissions_from_Indian_Livestock_in_Comparisons_with_GHG_Emissions_from_World_and_Developing_Countries, accessed on 28 October 2022

35 '20th Livestock Census', Vikaspedia, https://vikaspedia.in/agriculture/agri-directory/reports-and-policy-briefs/20th-livestock-census, accessed on 28 October 2022

36 'Poultry farming, climate change, and drivers of antimicrobial resistance in India', ResearchGate, https://www.researchgate.net/publication/338070789_Poultry_farming_climate_change_and_drivers_of_antimicrobial_resistance_in_India, accessed on 28 October 2022

37 Maneka Gandhi, 'Quality of egg is an issue', *The Statesman*, https://www.thestatesman.com/supplements/8thday/quality-egg-issue-1502756802.html, accessed on 28 October 2022

38 'A study on the quality of shelled eggs marketed in and around Hyderabad', India—IVORY Research, https://www.ivoryresearch.com/samples/a-study-on-the-quality-of-shelled-eggs-marketed-in-and-around-hyderabad-india/, accessed on 28 October 2022

39 'How Animal Agriculture Contributes to Water Shortages, Climate Change', Peta India, https://www.petaindia.com/blog/how-animal-agriculture-contributes-to-water-shortages-climate-change, accessed on 28 October 2022

40 Corrigendum to 'Greenhouse gas emissions from agricultural food production to supply Indian diets: Implications for climate change mitigation', Agriculture, Ecosystems and Environment, ScienceDirect, https://www.sciencedirect.com/science/article/pii/S0167880918304687, visited on 28 October 2022

41 'Preventing the Next Pandemic Zoonotic diseases and how to break the chain of transmission', https://www.cbd.int/doc/c/084c/e8fd/84ca7fe0e19e69967bb9fb73/unep-sa-sbstta-sbi-02-en.pdf, accessed on 28 October 2022

42 *Seaspiracy* directed by Ali Tabrizi, https://www.seaspiracy.org, accessed on 28 October 2022

43 'Study finds hazardous levels of insecticides, antibiotics in fish, shrimp farming in 10 states', FIAPO, https://www.fiapo.org/fiaporg/news/study-finds-hazardous-levels-of-insecticides-antibiotics-in-fish-shrimp-farming-in-10-states/, accessed on 28 October 2022

44 Study finds metal pollution in aquaculture farms', FIAPO, https://www.fiapo.org/fiaporg/news/study-finds-metal-pollution-in-aquaculture-farms/, accessed on 28 October 2022

45 'Toxicity in fish polluted with heavy metals, chemicals or drugs', ResearchGate, https://www.researchgate.net/publication/270213954_Toxicity_in_fish_polluted_with_heavy_metals_chemicals_or_drugs, accessed on 28 October 2022

46 'Plasticenta: First evidence of microplastics in human placenta', Environment International—Science Direct, https://www.sciencedirect.com/science/article/pii/S0160412020322297, accessed on 28 October 2022

47 'Balancing the Global Carbon Budget', Annual Review of Earth and Planetary Sciences, https://www.annualreviews.org/doi/abs/10.1146/annurev.earth.35.031306.140057, accessed on 28 October 2022

48 'Time to Get Real. There Aren't Plenty of Fish in the Sea and It's Our Fault', One Green Planet, https://www.onegreenplanet.org/environment/there-arent-plenty-of-fish-in-the-sea/, accessed on 28 October 2022

49 'Water used per litre of product produced by the Coca-Cola system from 2012 to 2021', Statista, https://www.statista.com/statistics/1234225/water-use-ratio-coca-cola-company-globally, accessed on 28 October 2022

50 'How Much Sugar is Actually in Coke – and what's the difference between Diet Coke and Coke Zero?', Good To Know, https://www.goodto.com/food/sugar-in-coke-524085, accessed on 28 October 2022

51 'Coca Cola Sales in Number of Unit Cases', Statista, https://statstic.com/coca-cola-sales-in-number-of-unit-cases, accessed on 28 October 2022

52 'Colonizing the Rains: Disentangling More-than-Human Technopolitics of Drought Protection in the Archive', Geoforum, ScienceDirect, https://www.sciencedirect.com/science/article/pii/S0016718522001464, accessed on 28 October 2020

53 'The Disturbing Secret Behind the World's Most Expensive Coffee', National Geographic, https://www.nationalgeographic.com/animals/article/160429-kopi-luwak-captive-civet-coffee-Indonesia, accessed on 28 October 2020

54 'We Have Exposed Sickening Animal Experiments', Cruelty Free International, https://crueltyfreeinternational.org/latest-news-and-updates/we-have-exposed-sickening-animal-experiments-danone-yakult-and-nestlé, visited on 28 October 2022

55 'The Growth Opportunities and Challenges for India's Frozen Food Market', India Retailing, https://www.indiaretailing.com/2022/03/07/food/the-growth-opportunities-of-indias-frozen-food-market, accessed on 28 October 2022

56 'Frozen, fresh or canned food: What's more nutritious?', BBC, https://www.bbc.com/future/article/20200427-frozen-fresh-or-canned-food-whats-more-nutritious, accessed on 28 October 2022

57 'India Frozen Food Market by Segments, End Users, Regions, Company Analysis, Forecast', Renub Research—Research and Markets, https://www.researchandmarkets.com/reports/5317023/india-frozen-food-market-by-segments-end-users, accessed on 28 October 2022

58 'How the World Got Hooked on Palm Oil', the *Guardian*, https://www.theguardian.com/news/2019/feb/19/palm-oil-ingredient-biscuits-shampoo-environmental, accessed on 28 October 2022

59 'Your Shampoo, Ice Cream, Cooking Oil and Lipstick—All Contain Palm Oil', WWF, https://www.wwfindia.org/about_wwf/making_businesses_sustainable/palm_oil/peel_back, accessed on 28 October 2022,

60 'Fertile Ground—India's Domestic Palm Oil Initiative is a Springboard for Sustainable Production', Roundtable on Sustainable Palm Oil, https://rspo.org/news-and-events/news/fertile-ground--indias-domestic-palm-oil-initiative-is-a-springboard-for-sustainable-production, visited on 28 October 2022

61 Just one hectare of forests cleared to grow oil palms is roughly equivalent to the amount of carbon produced by 530 people flying from Geneva to New York in economy class. The distance is approximately 6212 km. A similar distance in an Indian context would be travelling to Srinagar from Chennai and back.

62 'Palm Oil and Biodiversity', International Union for Conservation of Nature—Governing body of protected sites, https://www.iucn.org/resources/issues-brief/palm-oil-and-biodiversity, visited on 28 October 2022,

63 'Pulling Water Out of the Ground May Lead to Quakes on the San Andreas Fault', Smithsonian Magazine, https://www.smithsonianmag.com/science-nature/pulling-water-ground-lead-quakes-san-andreas-fault-180951456, accessed on 28 October 2022

64 'Food Carbon Footprint Calculator', My Emissions, https://myemissions.green/food-carbon-footprint-calculator, accessed on 28 October 2022

65 'CUFR Tree Carbon Calculator (CTCC)', USDA, https://www.fs.usda.gov/ccrc/tool/cufr-tree-carbon-calculator-ctcc, accessed on 28 October 2022

66 'Carbon footprints of Indian food items', ResearchGate, https://www.researchgate.net/publication/223499350_

Carbon_footprints_of_Indian_food_items, accessed on 28 October 2022

67 Daily Dump, visited on 28 October 2022, https://www.dailydump.org

68 'How cutting your food waste can help the climate', BBC, visited on 28 October 2022, https://www.bbc.com/future/article/20200224-how-cutting-your-food-waste-can-help-the-climate

69 'Meat consumption', Australian Government Department of Agriculture, Fisheries and Forestry, https://www.agriculture.gov.au/abares/research-topics/agricultural-outlook/meat-consumption, accessed on 28 October 2022

70 'Food Intake Biomarkers for Green Leafy Vegetables, Bulb Vegetables, and Stem Vegetables: A Review', National Institute of Health—National Library of Medicine—National Center for Biotechnology Information, https://www.ncbi.nlm.nih.gov/pmc/articles/PMC7144047, accessed on 28 October 2022

71 'The Future of Food is Biodiverse', Future Market, https://thefuturemarket.com/biodiversity, accessed on 28 October 2022

72 'The Mushroom Sustainability Story: Water, Energy, and Climate Environmental Metrics', Sure Harvest, https://www.mushroomcouncil.com/wp-content/uploads/2017/12/Mushroom-Sustainability-Story-2017.pdf, accessed on 28 October 2022

73 'Sustainability', https://www.americanmushroom.org/main/sustainability, accessed on 28 October 2022

74 '50 foods for healthier people and a healthier planet', WWF, https://www.wwf.org.uk/sites/default/files/2019-02/Knorr_Future_50_Report_FINAL_Online.pdf, accessed on 28 October 2022

75 'Beef: The "King" of the Big Water Footprints', Water Footprint Calculator, https://www.watercalculator.org/news/articles/beef-king-big-water-footprints, accessed on 28 October 2022

76 'Nutraceutical Value of Finger Millet [Eleusine coracana (L.) Gaertn.], and Their Improvement Using Omics Approaches', Frontiers in Plant Science—ResearchGate, https://www.researchgate.net/publication/304784794_Nutraceutical_Value_of_Finger_Millet_Eleusine_coracana_L_Gaertn_and_Their_Improvement_Using_Omics_Approaches, accessed on 28 October 2022

Chapter 3

1   'Measuring Carbon Emissions in the Garment Sector in Asia', International Labour Organization, https://www.ilo.org/legacy/english/intserv/working-papers/wp053/index.html, accessed on 28 October 2022
2   'Textiles and Apparel sixth largest exporter of Textiles and Apparel', Invest India, visited on 28 October 2022, https://www.investindia.gov.in/sector/textiles-apparel
3   'Textile Market Size, Share and Trends Analysis Report By Raw Material (Cotton, Wool, Silk, Chemical), By Product (Natural Fibers, Nylon), By Application (Technical, Fashion), By Region, And Segment Forecasts, 2022–2030', Grand View Research, visited on 28 October 2022, https://www.grandviewresearch.com/industry-analysis/textile-market
4   'Textile Industry and Market Growth in India', India Brand Equity Foundation, visited on 28 October 2022, https://www.ibef.org/industry/textiles
5   'Value of textile imported into India from financial year 2011 to 2022', Statista, visited on 28 October 2022, https://www.statista.com/statistics/625178/import-value-of-textile-india
6   'Putting the brakes on fast fashion', United Nations Environment Programme, visited on 28 October 2022, https://www.unep.org/news-and-stories/story/putting-brakes-fast-fashion

7   'The price of fast fashion', Nature Climate Change, visited on 28 October 2022, https://www.nature.com/articles/s41558-017-0058-9

8   'Measuring carbon emissions in the garment sector in Asia', International Labour Organization, https://www.ilo.org/legacy/english/intserv/working-papers/wp053/index.html, accessed on 28 October 2022

9   'The Apparel Industry's Environmental Impact in Six Graphics', World Resource Institute, https://www.wri.org/insights/apparel-industrys-environmental-impact-6-graphics, accessed on 28 October 2022

10   T. Karthik and D. Gopalakrishnan, 'Environmental Analysis of Textile Value Chain: An Overview. Roadmap to Sustainable Textiles and Clothing', 2014, pp. 153-188, https://www.researchgate.net/publication/281104248_Environmental_Analysis_of_Textile_Value_Chain_An_Overview, accessed on 28 October 2022

11   'EP-Textiles Coordination Division', Ministry of Commerce and Industry, https://commerce.gov.in/about-us/divisions/export-products-division/ep-textile, accessed on 28 October 2022

12   T. Karthik and D. Gopalakrishnan, 'Environmental Analysis of Textile Value Chain: An Overview. Roadmap to Sustainable Textiles and Clothing', 2014, pp. 153-188, https://www.researchgate.net/publication/281104248_Environmental_Analysis_of_Textile_Value_Chain_An_Overview, accessed on 28 October 2022

13   'Cotton Fibre – Tribology of Natural Fiber Polymer Composites (Second Edition), 2021', Science Direct, https://www.sciencedirect.com/topics/engineering/cotton-fibre, accessed on 28 October 2022

14   'Animals Used for Clothing', PETA, https://www.peta.org/issues/animals-used-for-clothing, accessed on 28 October 2022

15   'Is the wool industry ethical?', Wool Facts, visited on 28 October 2022, https://www.woolfacts.com, accessed on 28 October 2022

16 'Wool', PETA India, visited on 28 October 2022, https://
    www.petaindia.com/issues/animals-used-for-clothing/
    wool, accessed on 28 October 2022

17 'Leather', PETA India, https://www.petaindia.com/
    issues/animals-used-for-clothing/leather, accessed on
    28 October 2022

18 'Fur Trapping', PETA, https://www.peta.org/issues/
    animals-used-for-clothing/fur/fur-trapping, accessed on
    28 October 2022

19 'Sustainable silk production - Sustainable Fibres and
    Textiles', The Textile Institute Book Series, visited on 28
    October 2022, https://www.sciencedirect.com/science/
    article/pii/B9780081020418000068

20 'Water footprint assessment of handwoven silk
    production', Wiley Online Library, visited on 28 October
    2022, https://onlinelibrary.wiley.com/doi/abs/10.1111/
    wej.12674

21 'The Environmental Price of Fast Fashion', Nature
    Reviews Earth & Environment volume 1, pp. 189–200
    (2020), https://www.nature.com/articles/s43017-020-
    0039-9, accessed on 28 October 2022

22 'The Environmental Price of Fast Fashion', Nature
    Reviews Earth & Environment volume 1, pp. 189–200
    (2020), https://www.nature.com/articles/s43017-020-
    0039-9, accessed on 28 October 2022

23 'Introduction to Sustainability and the Textile Supply
    Chain and its Environmental Impact—Assessing the
    Environmental Impact of Textiles and the Clothing
    Supply Chain', ResearchGate, pp.1-32, https://www.
    researchgate.net/publication/340242239_Introduction_
    to_sustainability_and_the_textile_supply_chain_and_
    its_environmental_impact, accessed on 28 October 2022

24 'What is Microfiber Plastic Pollution', Save Our Shores,
    https://saveourshores.org/microfibers, accessed on 28
    October 2022

25 Róisín Magee Altreuter, 'Microfibers, Macro problems—A
    resource guide and toolkit for understanding and

tackling the problem of plastic microfiber pollution in our communities', https://static1.squarespace.com/static/5522e85be4b0b65a7c78ac96/t/5a66456cc83025f49135fbc2/1516651904354/Microfibers%2C+Macro+problems.pdf<, accessed on 28 October 2022

26 'Accumulations of Microplastic on Shorelines Worldwide: Sources and Sinks', https://www.plasticsoupfoundation.org/wp-content/uploads/2015/03/Browne_2011-EST-Accumulation_of_microplastics-worldwide-sources-sinks.pdf, accessed on 28 October 2022

27 'Does Use Matter? Comparison of Environmental Impacts of Clothing Based on Fiber Type', https://www.mdpi.com/2071-1050/10/7/2524/pdf, accessed on 28 October 2022

28 'The Environmental Price of Fast Fashion', Nature Reviews Earth & Environment volume 1, pp. 189–200 (2020), https://www.nature.com/articles/s43017-020-0039-9, accessed on 28 October 2022

29 'How Dialogue is Shifting Bangladesh's Textile Industry from Pollution Problem to Pollution Solution', The World Bank, https://www.worldbank.org/en/news/feature/2017/02/15/how-dialogue-is-shifting-bangladeshs-textile-industry-from-pollution-problem-to-pollution-solution, accessed on 28 October 2022

30 'Cleaning up China's Polluted Pearl River', The World Bank, https://www.worldbank.org/en/results/2016/05/26/cleaning-up-china-polluted-pearl-river, accessed on 28 October 2022

31 'Textile Dyeing Industry an Environmental Hazard', Scientific Research, https://www.scirp.org/journal/paperinformation.aspx, accessed on 28 October 2022

32 'Toxic Threads: Polluting Paradise', GreenPeace, visited on 28 October 2022, https://www.greenpeace.org/static/planet4-international-stateless/2013/04/62ec9171-toxic-threads-04.pdf

33 'Marking Progress Against Child Labour Global Estimates and Trends 2000-2012', International Labour Office, visited on 28 October 2022, https://www.ilo.org/wcmsp5/groups/public/---ed_norm/---ipec/documents/publication/wcms_221513.pdf, accessed on 28 October 2022

34 'Sustainable Clothing Action Plan', DEFRA GOV UK, https://assets.publishing.service.gov.uk/government/uploads/system/uploads/attachment_data/file/69193/pb13206-clothing-action-plan-100216.pdf, accessed on 28 October 2022

35 'Marking Progress Against Child Labour', International Labour Office, visited on 28 October 2022, https://www.ilo.org/wcmsp5/groups/public/---ed_norm/---ipec/documents/publication/wcms_221513.pdf, accessed on 28 October 2022

36 'The Apparel Industry's Environmental Impact in Six Graphics', World Resource Institute, visited on 28 on 2022, https://www.wri.org/insights/apparel-industrys-environmental-impact-6-graphics

37 'The US Fashion Report', India Retailing, https://www.indiaretailing.com/wp-content/uploads/dlm_uploads/2018/08/US-Fashion-Report-2018-Final.pdf, accessed on 28 October 2022

38 'Apparent consumption of apparel worldwide from 2013 to 2026', Statista, https://www.statista.com/forecasts/1307850/worldwide-consumption-of-clothing-items, accessed on 28 October 2022

39 'A New Textiles Economy: Redesigning fashion's future', Ellen Macarthur Foundation, https://ellenmacarthurfoundation.org/a-new-textiles-economy, accessed on 28 October 2022

40 'The price of fast fashion', Nature Climate Change, https://www.nature.com/articles/s41558-017-0058-9, accessed on 28 October 2022

41 'Fashion on Climate', Global Fashion of Agenda, https://globalfashionagenda.org/resource/fashion-on-climate, accessed on 28 October 2022

42 'Where Do Our Unwanted Clothes Go?', Cooper Hewitt, https://www.cooperhewitt.org/2016/12/20/infographic-where-do-our-unwanted-clothes-go, accessed on 28 October 2022

43 'Poisoned Gifts - From Donations to the Dumpsite', GreenPeace, https://www.greenpeace.org/static/planet4-international-stateless/2022/04/9f50d3de-greenpeace-germany-poisoned-fast-fashion-briefing-factsheet-april-2022.pdf, accessed on 28 October 2022

44 'Poisoned Gifts: From Donations to the Dumpsite', GreenPeace, https://www.greenpeace.org/static/planet4-international-stateless/2022/04/9f50d3de-greenpeace-germany-poisoned-fast-fashion-briefing-factsheet-april-2022.pdf, accessed on 28 October 2022

45 'Fashion on Climate', Global Fashion Agenda, https://globalfashionagenda.org/resource/fashion-on-climate, accessed on 28 October 2022

46 'Fashion on Climate', Global Fashion Agenda, https://globalfashionagenda.org/resource/fashion-on-climate, accessed on 28 October 2022

47 'HEMP', The Council of Fashion Designers of America, Inc., https://cfda.com/resources/materials/detail/hemp, accessed on 28 October 2022

48 'Is Bamboo Fabric as Eco-friendly as Bamboo?', Fiber 2 Fashion, https://www.fibre2fashion.com/industry-article/7347/is-bamboo-fabric-as-eco-friendly-as-bamboo, accessed on 28 October 2022

49 'An Assessment of Environment Friendly Methods of Khadi Manufacturing', Indian Journal of History of Science, Springer Link, https://link.springer.com/article/10.1007/s43539-021-00003-3, accessed on 28 October 2022

50 'The UK Ethical Market', Ethical Consumer, https://www.ethicalconsumer.org/retailers/uk-ethical-market, accessed on 28 October 2022

51 'The Indian Startup That's Turning Air Pollution Into Ink', https://homegrown.co.in/homegrown-voices/the-delhi-

startup-thats-turning-air-pollution-into-ink, accessed on 28 October 2022

52 'Shop Less, Mend More: Making More Sustainable Fashion Choices', the *Guardian*, https://www.theguardian.com/lifeandstyle/2018/feb/10/shop-less-mend-more-making-more-sustainable-fashion-choices, accessed on 28 October 2022

53 'Price List', GoodWill, https://goodwillsp.org/shop/price-list, accessed on 28 October 2022

54 'Women's Blazers', *Express*, https://www.express.com/womens-clothing/tops/blazers/cat1910071, accessed on 28 October 2022

55 '20 Hard Facts and Statistics About Fast Fashion', Good On You, https://goodonyou.eco/fast-fashion-facts, accessed on 28 October 2022

56 'Valuing our Clothes: The Cost of UK Fashion', https://wrap.org.uk/resources/report/valuing-our-clothes-cost-uk-fashion, accessed on 28 October 2022

57 'Fashion on Climate', https://globalfashionagenda.org/resource/fashion-on-climate, accessed on 28 October 2022

58 'Dropping Centers', Goonj, https://goonj.org/dropping-centres, accessed on 28 October 2022

59 'Collection Camps', Goonj, https://goonj.org/collection-camps, accessed on 28 October 2022

60 'What's the Carbon Footprint of . . . a Load of Laundry?', the *Guardian*, https://www.theguardian.com/environment/green-living-blog/2010/nov/25/carbon-footprint-load-laundry, accessed on 28 October 2022

61 'A New Textiles Economy: Redesigning Fashion's Future', Ellen Macarthur Foundation, https://ellenmacarthurfoundation.org/a-new-textiles-economy, accessed on 28 October 2022

62 'Fast fashion: Inside the fight to end the silence on waste', BBC, https://www.bbc.com/news/world-44968561, accessed on 28 October 2022

Chapter 4

1   'Cars, Planes, Trains: Where do CO2 Emissions from Transport Come From?', Our World in Data, https://ourworldindata.org/co2-emissions-from-transport, accessed on 28 October 2022

2   'Greenhouse Gas Equivalencies Calculator', United States Environmental Protection Agency, https://www.epa.gov/energy/greenhouse-gas-equivalencies-calculator#results, accessed on 28 October 2022

3   'Profiles – India', OEC, https://oec.world/en/profile/country/ind, accessed on 28 October 2022

4   'The Scale and Drivers of Carbon Footprints in Households, Cities and Regions Across India - Global Environment Change', ScienceDirect, https://www.sciencedirect.com/science/article/pii/S0959378020307883, accessed on 28 October 2022

5   'INDIA'S MERCHANDISE TRADE: Preliminary Data of May 2022', Ministry of Commerce and Industry, https://pib.gov.in/PressReleasePage.aspx, accessed on 28 October 2022

6   'Logistics workers: Heroes in the shadows', Global Logistics Network, https://www.globalialogisticsnetwork.com/blog/2020/05/21/logistics-workers-heroes-in-the-shadows, accessed on 28 October 2022

7   'Cars, Planes, Trains: Where do CO2 Emissions from Transport Come From?', Our World in Data, https://ourworldindata.org/co2-emissions-from-transport, accessed on 28 October 2022

8   Ibid

9   'Transportation and Climate Change | Carbon Emissions by Transport Type', Blog Mech, https://blogmech.com/transportation-and-climate-change, accessed on 28 October 2022

10   'Global Greenhouse Gas Emissions by the Transportation Sector', The Geography of Transport Systems, https://transportgeography.org/contents/chapter4/

transportation-and-environment/greenhouse-gas-emissions-transportation, accessed on 28 October 2022

11 'The Top Causes Of Excess GHG Emissions In Transportation', Mercury Gate, https://mercurygate.com/collateral/infographic/the-top-causes-of-excess-ghg-emissions-in-transportation, accessed on 28 October 2022

12 'Electric Vehicles: World Energy Outlook', visited on 28 October 2022, https://www.iea.org/reports/electric-vehicles

13 'Which Form of Transport Has The Smallest Carbon Footprint?', Our World in Data, https://ourworldindata.org/travel-carbon-footprint, accessed on 28 October 2022

14 'Should We Give Up Flying for the Sake Of The Climate?', BBC, https://www.bbc.com/future/article/20200218-climate-change-how-to-cut-your-carbon-emissions-when-flying, accessed on 28 October 2022

15 'Guest post: Calculating the true climate impact of aviation emissions', CarbonBrief, https://www.carbonbrief.org/guest-post-calculating-the-true-climate-impact-of-aviation-emissions, accessed on 28 October 2022

16 'The Environmental Effects of Freight', OECD, https://www.oecd.org/environment/envtrade/2386636.pdf, accessed on 28 October 2022

17 'Shipping and World Trade: Driving Prosperity', International Chamber of Shipping, https://www.ics-shipping.org/shipping-fact/shipping-and-world-trade-driving-prosperity, accessed on 28 October 2022

18 'Chapter 27 - Environmental Effects of Marine Transportation', World Seas: An Environmental Evaluation (Second Edition), ScienceDirect, https://www.sciencedirect.com/science/article/pii/B9780128050521000309, accessed on 28 October 2022

19 'Follow the Friendly Floatees', National Geographic Society, https://www.nationalgeographic.org/activity/follow-friendly-floatees, accessed on 28 October 2022

20 'Leaky Gas Pipeline Sparks an Inferno in the Gulf of Mexico', *New York Times*, https://www.nytimes.com/2021/07/03/world/americas/eye-fire-gulf-mexico.html, accessed on 28 October 2022

21 'A Review on the Environmental Impacts of Shipping on Aquatic and Nearshore Ecosystems', Science of the Total Environment, ScienceDirect, accessed on 28 October 2022 https://www.sciencedirect.com/science/article/abs/pii/S0048969719335624

22 'Reduction of Maritime GHG Emissions and the Potential Role of E-fuels', Transportation Research Part D: Transport and Environment, ScienceDirect, https://www.sciencedirect.com/science/article/pii/S1361920921003722, accessed on 28 October 2022

23 'Greenhouse Gas Emissions from a Typical Passenger Vehicle', United States Environmental Protection Agency, https://www.epa.gov/greenvehicles/greenhouse-gas-emissions-typical-passenger-vehicle, accessed on 28 October 2022

24 'Road Freight Fleet Emissions Worldwide as of 2020, by Vehicle Type', Stastista, https://www.statista.com/statistics/1200116/road-freight-emissions-by-vehicle-type-worldwide, accessed on 28 October 2022

25 'Road Freight Zero: Pathways to Faster Adoption of Zero-Emission Trucks', McKinsey & Company, https://www.mckinsey.com/~/media/mckinsey/industries/automotive%20and%20assembly/our%20insights/road%20freight%20global%20pathways%20report/pathways-to-faster-adoption-of-zero-emission-trucks-2021.pdf, accessed on 28 October 2022

26 'Toy imports down by 70% and exports up 61% over last three years as Make in India yields positive results for the sector', Ministry of Commerce and Industry, https://pib.gov.in/PressReleaseIframePage.aspx, accessed on 28 October 2022

27 'The Ecological Footprint of Ecotourism Packages', Global Footprint Network, https://www.footprintnetwork.org/

our-work/sustainable-tourism, accessed on 28 October 2022

28 'Automobile Domestic Sales Trends', Society of Indian Automobile Manufacturers, https://www.siam.in/ statistics.aspx?mpgid=8&pgidtrail=14, accessed on 28 October 2022

29 'Average Daily Ridership of Public Road Transport in Mumbai, India from Financial Year 2010 to 2021', Statista, https://www.statista.com/statistics/1240100/india-average-daily-ridership-of-buses-in-mumbai, accessed on 28 October 2022

30 'Compare and Book Flights and we'll Offset your Flight's Carbon Emissions for Free', FlyGreen, https://flygrn. com/#how, accessed on 28 October 2022

Chapter 5

1 'Carbon Footprint of Tourism', Sustainable Travel International, https://sustainabletravel.org/issues/ carbon-footprint-tourism, accessed on 28 October 2022

2 'Current Global Tourism', Gujarat Infrastructure Development Board, https://www.gidb.org/tourism-current-global-tourism, accessed on 28 October 2022

3 'Total Contribution of Travel and Tourism to Gross Domestic Product (GDP) Worldwide from 2019 to 2021', Statista, https://www.statista.com/statistics/233223/ travel-and-tourism--total-economic-contribution-worldwide, accessed on 28 October 2022

4 'The Carbon Footprint of Global Tourism', Nature Climate Change, https://www.nature.com/articles/s41558-018-0141-x, accessed on 28 October 2022

5 'Carbon Footprint of Tourism', Sustainable Travel International, https://sustainabletravel.org/issues/ carbon-footprint-tourism, accessed on 28 October 2022

6 'Visualizing the Countries Most Reliant on Tourism', Visual Capitalist, https://www.visualcapitalist.com/ countries-reliant-tourism, accessed on 28 October 2022

7   'Constructing an "Iron" Unity: The Statue of Unity and India's Nationalist Historiography', Australian Journal of Politics and History: Volume 66, Number 2, 2020, https://doi.org/10.1111/ajph.12678, accessed on 28 October 2022

8   'Tourism Boosts Human Trafficking in Goa', *Times of India*, https://timesofindia.indiatimes.com/city/goa/Tourism-boosts-human-trafficking-in-Goa/articleshow/21563733.cms, accessed on 28 October 2022

9   'The carbon footprint of global tourism', Nature Climate Change, https://www.nature.com/articles/s41558-018-0141-x, accessed on 28 October 2022

10 Climate Change Everyone's Business—Implications of Tourism, Tourism on the Move in a Changing Climate, University of Cambridge, Cambridge Judge Business School, Cambridge Institute for Sustainability Leadership, https://www.cisl.cam.ac.uk/system/files/documents/ipcc-ar5-implications-for-tourism-briefing-prin.pdf, accessed on 28 October 2022

11 'Wildlife is Our World Heritage', World Tourism Organization, UNWTO, https://www.unwto.org/asia/unwto-chimelong-why-wildlife, accessed on 28 October 2022

12 'Wildlife Tourists in India's Emerging Economy: Potential for a Conservation Constituency?', ResearchGate, https://www.researchgate.net/publication/259424331_Wildlife_tourists_in_India's_emerging_economy_Potential_for_a_conservation_constituency, accessed on 28 October 2022

13 '5 Things *Tiger King* Doesn't Explain About Captive Tigers', WWF, https://www.worldwildlife.org/stories/5-things-tiger-king-doesn-t-explain-about-captive-tigers, accessed on 28 October 2022

14 'Facts', WWF, https://www.worldwildlife.org/species/tiger, accessed on 28 October 2022

15 'Owning Wild Animals: Stats on Exotic Pets', Live Science, https://www.livescience.com/16815-exotic-

pets-wildlife-infographic.html, accessed on 28 October 2022

16 'New research: Calculating the Carbon Emissions of Holidays', Responsible Travel, https://www. responsibletravel.com/copy/carbon-emissions-of-holidays, accessed on 28 October 2022

17 'Carbon Footprint of Travel Per Kilometer 2018', Our World in Data, https://ourworldindata.org/grapher/carbon-footprint-travel-mode, accessed on 28 October 2022

18 'Should We Give Up Flying for the Sake of the Climate?', BBC, https://www.bbc.com/future/article/20200218-climate-change-how-to-cut-your-carbon-emissions-when-flying, accessed on 28 October 2022

19 'Small Percentage of Frequent Flyers Are Driving Global Emissions, New Study Shows', EcoWatch, https://www. ecowatch.com/frequent-flyer-emissions-2651292287. html, accessed on 28 October 2022

20 'Elite Minority of Frequent Flyers "cause most of aviation's climate damage", *Guardian*, https://www.theguardian. com/world/2021/mar/31/elite-minority-frequent-flyers-aviation-climate-damage-flights-environmental, accessed on 28 October 2022

21 'Aviation Climate Change Impact Infographic', Eureka Alert, https://www.eurekalert.org/multimedia/645450, accessed on 28 October 2022

22 'Climate Change and its Impacts on Tourism', WWF, https://assets.wwf.org.uk/downloads/tourism_and_cc_full.pdf, accessed on 28 October 2022

23 'Revenue of the Cruise Industry Worldwide from 2017 to 2026', Statista, https://www.statista.com/forecasts/1258061/revenue-cruises-worldwide, accessed on 28 October 2022

24 'The Environmental Challenges of Cruise Tourism: Impacts and Governance', ResearchGate, https:// www.researchgate.net/publication/286937048_The_environmental_challenges_of_cruise_tourism_impacts_and_governance, accessed on 28 October 2022

25 'Cruise Tourism in India: Sailing into Troubled Waters',
     Down To Earth, https://www.downtoearth.org.in/blog/
     climate-change/cruise-tourism-in-india-sailing-into-
     troubled-waters-75995, accessed on 28 October 2022
26 '7 Places Being Ruined by Cruise Ships', Insider, https://
     www.insider.com/cruise-ships-environmental-impact-
     tourism-2019-9, accessed on 28 October 2022
27 'The Environmental Challenges of Cruise Tourism:
     impacts and governance', ResearchGate, https://
     www.researchgate.net/publication/286937048_The_
     environmental_challenges_of_cruise_tourism_impacts_
     and_governance, accessed on 28 October 2022
28 'The Shipping Sector is Finally on Board in the Fight
     Against Climate Change', The Conversation, https://
     theconversation.com/the-shipping-sector-is-finally-
     on-board-in-the-fight-against-climate-change-95212,
     accessed on 28 October 2022
29 'The Environmental Challenges of Cruise Tourism:
     Impacts and Governance', ResearchGate, https://
     www.researchgate.net/publication/286937048_The_
     environmental_challenges_of_cruise_tourism_impacts_
     and_governance, accessed on 28 October 2022
30 'Cruise Tourism in India: Sailing into Troubled Waters',
     Down To Earth, https://www.downtoearth.org.in/blog/
     climate-change/cruise-tourism-in-india-sailing-into-
     troubled-waters-75995, accessed on 28 October 2022
31 'Vehicle Sharing—the Solution to Hazardous Air Pollution
     in India?' *Financial Express*, https://www.financialexpress.
     com/auto/car-news/vehicle-sharing-the-solution-to-
     hazardous-air-pollution-in-india/1494231, accessed on
     28 October 2022
32 'The Ecological Footprint of Ecotourism Packages',
     Global Footprint Network, visited on 28 October
     2022, https://www.footprintnetwork.org/our-work/
     sustainable-tourism
33 'The Ecological Footprint of Ecotourism Packages', Global
     Footprint Network, https://www.footprintnetwork.org/

our-work/sustainable-tourism, accessed on 28 October 2022

34 'Travel and Tourism Industry in India—statistics and facts', Statista, https://www.statista.com/topics/2076/travel-and-tourism-industry-in-india, accessed on 28 October 2022

35 'The Ecological Footprint of Ecotourism Packages', Global Footprint Network, https://www.footprintnetwork.org/our-work/sustainable-tourism, accessed on 28 October 2022

36 'Tourism, as it Is Now, Is Unsustainable', CondeNast Traveller, https://www.cntraveller.in/magazine-story/tourism-now-unsustainable, accessed on 28 October 2022

Chapter 6

1 'Towards Digital Sobriety', The Shift Project, https://theshiftproject.org/wp-content/uploads/2019/03/Lean-ICT-Report_The-Shift-Project_2019.pdf, accessed on 28 October 2022

2 'Information Technology Global Market Report 2022', The Business Research Company, \https://www.thebusinessresearchcompany.com/report/information-technology-global-market-report, accessed on 28 October 2022

3 'The Tech Workforce Is Expanding—and Changing—as Different Sectors Battle for Talent', Deloitte, https://www2.deloitte.com/us/en/insights/economy/spotlight/tech-workforce-expanding.html, accessed on 28 October 2022

4 'Land, Waste, and Cleanup Topics', United States Environmental Protection Agency, https://www.epa.gov/environmental-topics/land-waste-and-cleanup-topics, accessed on 28 October 2022

5 'The Carbon Impact of Streaming: An Update on BBC TV's Energy Footprint', BBC, https://www.bbc.

co.uk/rd/blog/2021-06-bbc-carbon-footprint-energy-envrionment-sustainability, accessed on 28 October 2022

6   'How Clean is Your Cloud?', Greenpeace, https://www.greenpeace.org/international/publication/6986/how-clean-is-your-cloud, accessed on 28 October 2022

7   'Achieving Our Goal: 100% Renewable Energy for our Global Operations', Meta, https://tech.fb.com/engineering/2021/04/renewable-energy, accessed on 28 October 2022

8   '100% Renewable Is Just the Beginning', Google, https://sustainability.google/progress/projects/announcement-100, accessed on 28 October 2022

9   'Apple Powers Ahead in New Renewable Energy Solutions with Over 110 Suppliers', Apple, https://www.apple.com/in/newsroom/2021/03/apple-powers-ahead-in-new-renewable-energy-solutions-with-over-110-suppliers, accessed on 28 October 2022

10 'Accelerating the Global Transition to Cleaner Energy', Microsoft, https://www.microsoft.com/en-us/sustainability/energy, accessed on 28 October 2022

11 'Where Are the World's Cloud Data Centers and Who is Using Them?', TeleGeography, https://blog.telegeography.com/where-are-the-worlds-cloud-data-centers-and-who-is-using-them, accessed on 28 October 2022

12 'Cooling Energy Consumption Investigation of Data Center IT Room with Vertical Placed Server', ResearchGate, https://www.researchgate.net/publication/317308758_Cooling_Energy_Consumption_Investigation_of_Data_Center_IT_Room_with_Vertical_Placed_Server, accessed on 28 October 2022

13 'The (More) Sustainable Data Center', Capgemini, https://www.capgemini.com/insights/expert-perspectives/the-more-sustainable-data-center, accessed on 28 October 2022

14 'Climate Change: Is Your Netflix Habit Bad for the Environment?', BBC, https://www.bbc.com/news/technology-45798523, accessed on 28 October 2022

15 'How to Stop Data Centres from Gobbling up the World's Electricity', Nature, https://www.nature.com/articles/d41586-018-06610-y, accessed on 28 October 2022

16 'Powering the Internet: Your Virtual Carbon Footprint', WebFX, https://www.webfx.com/blog/marketing/carbon-footprint-internet, accessed on 28 October 2022

17 'Our Digital Carbon Footprint: What's the Environmental Impact of the Online World?', Reset, https://en.reset.org/knowledge/our-digital-carbon-footprint-whats-the-environmental-impact-online-world-12302019, accessed on 28 October 2022

18 'Individuals Using the Internet (% of population)—India', The World Bank, 2022, https://data.worldbank.org/indicator/IT.NET.USER.ZS?end=2020&locations=IN&start=2000&view=chart, accessed on 28 October 2022

19 'Clicking Clean: Who Is Winning the Race to Build a Green Internet?', Greenpeace, https://www.greenpeace.org/static/planet4-international-stateless/2017/01/35f0ac1a-clickclean2016-hires.pdf, accessed on 28 October 2022

20 Ibid

21 Ibid

22 'Clicking clean: Who Is Winning the Race to Build a Green Internet?', GreenPeace, https://www.greenpeace.org/static/planet4-international-stateless/2017/01/35f0ac1a-clickclean2016-hires.pdf, accessed on 28 October 2022

23 'Data, Digital Technology, and the Environment', Geneva Environment Network, https://www.genevaenvironmentnetwork.org/resources/updates/data-digital-technology-and-the-environment, accessed on 28 October 2022

24 'Number of Digital Video Viewers Worldwide from 2019 to 2023', Statista, https://www.statista.com/statistics/1061017/digital-video-viewers-number-worldwide, accessed on 28 October 2022

25 'Global Video Streaming Market Share Report, 2030', Grand View Research, https://www.grandviewresearch.com/industry-analysis/video-streaming-market, accessed on 28 October 2022

26 'Impacts of the Digital Transformation on the Environment and Sustainability', ResearchGate, https://www.researchgate.net/publication/342039732_Impacts_of_the_digital_transformation_on_the_environment_and_sustainability, accessed on 28 October 2022

27 'The Carbon Footprint of Streaming Video: Fact-checking the Headlines', World Energy Outlook, https://www.iea.org/commentaries/the-carbon-footprint-of-streaming-video-fact-checking-the-headlines, accessed on 28 October 2022

28 'Climate Crisis: The Unsustainable use of Online Video', The Shift Project, https://theshiftproject.org/wp-content/uploads/2019/07/2019-02.pdf, accessed on 28 October 2022

29 'A Quick Guide to Your Digital Carbon Footprint', Ericsson, https://www.ericsson.com/en/reports-and-papers/industrylab/reports/a-quick-guide-to-your-digital-carbon-footprint, accessed on 28 October 2022

30 'Climate Crisis: The Unsustainable use of Online Video', The Shift Project, https://theshiftproject.org/wp-content/uploads/2019/07/2019-02.pdf, accessed on 28 October 2022

31 'ICT and the Climate', Ericsson, https://www.ericsson.com/en/reports-and-papers/industrylab/reports/a-quick-guide-to-your-digital-carbon-footprint, accessed on 28 October 2022

32 'Clicking clean: Who Is Winning the Race to Build a Green Internet?', Greenpeace, visited on 28 October 2022, https://www.greenpeace.org/static/

planet4-international-stateless/2017/01/35f0ac1a-clickclean2016-hires.pdf, accessed on 28 October 2022

33 S. Frankel and D. Gervais, eds., *The Evolution and Equilibrium of Copyright in the Digital Age*, volume 26 (Cambridge University Press, 2014)

34 'Streaming Music Is Driving Up Harmful Emissions, According to Study', Fact Mag, https://www.factmag.com/2019/04/09/streaming-music-emissions-study, accessed on 28 October 2022

35 'Last Night a DJ Took a Flight', CleanScene, https://cleanscene.club/report.pdf, accessed on 28 October 2022

36 'Clicking clean: Who is winning the race to build a green internet?', Greenpeace, https://www.greenpeace.org/static/planet4-international-stateless/2017/01/35f0ac1a-clickclean2016-hires.pdf, accessed on 28 October 2022

37 'The Environmental Cost Of Your Internet Searches, Visualized', Fast Company, https://www.fastcompany.com/90171268/internet_impact_visualized, accessed 28 October 2022

38 'The Environmental Impact of Search Engines Apps', Greenspector, https://greenspector.com/en/search-engines, accessed on 28 October 2022

39 Ibid

40 'Powering a Google search', Google, https://googleblog.blogspot.com/2009/01/powering-google-search.html, accessed on 28 October 2022

41 'Google Sustainability', Google, https://sustainability.google/reports/environmental-report-2019/#data-centers, accessed on 28 October 2022

42 'Microsoft Will Be Carbon Negative by 2030', Microsoft, https://blogs.microsoft.com/blog/2020/01/16/microsoft-will-be-carbon-negative-by-2030, accessed on 28 October 2022

43 'Clicking Clean: Who Is Winning the Race to Build a Green Internet?', Greenpeace, https://www.

greenpeace.org/static/planet4-international-stateless/2017/01/35f0ac1a-clickclean2016-hires.pdf, accessed on 28 October 2022

44 'How Bad Are Bananas?: The Carbon Footprint of everything', Profile Books, https://books.google.co.in/books?id=zs13m5JquBwC&newbks=0&hl=en&source=newbks_fb&redir_esc=y, accessed on 28 October 2022

45 'Why Your Internet Habits Are Not as Clean as You Think', BBC, https://www.bbc.com/future/article/20200305-why-your-internet-habits-are-not-as-clean-as-you-think, accessed on 28 October 2022

46 'Clicking Clean: Who Is Winning the Race to Build a Green Internet?', Greenpeace, https://www.greenpeace.org/static/planet4-international-stateless/2017/01/35f0ac1a-clickclean2016-hires.pdf, accessed on 28 October 2022

47 'The Carbon Impact of Instagram App Features', Greenspector, https://greenspector.com/en/6168-2, accessed on 28 October 2022

48 'Minute on the Internet in 2021', Statista, https://www.statista.com/chart/25443/estimated-amount-of-data-created-on-the-internet-in-one-minute, accessed on 28 October 2022

49 'Social Network Usage and Growth Statistics: How Many People Use Social Media in 2022?', BackLinkO, visited on 28 October 2022, https://backlinko.com/social-media-users

50 'The Environmental Impact of Amazon's Kindle', CleanTech Group LLC, 2022, https://gato-docs.its.txst.edu/jcr:4646e321-9a29-41e5-880d-4c5ffe69e03e/thoughts_ereaders.pdf, accessed on 28 October 2022

51 G.L. Kozak and G.A. Keolelan, 'Printed Scholarly Books and e-book Reading Devices: A Comparative Life Cycle Assessment of Two Book Options' (at IEEE International Symposium on Electronics and the Environment, May 2003), pp. 291-296

52 'E-Books Still No Match for Printed Books', Statista, https://www.statista.com/chart/24709/e-book-and-printed-book-penetration, accessed on 28 October 2022,

53 'At 30% CAGR, CryptoCurrency Market Cap Size Value Surges to Record $5,190.62 Million by 2026, Says Facts and Factors', Global Wisewire, https://www.globenewswire.com/en/news-release/2021/04/12/2208331/0/en/At-30-CAGR-CryptoCurrency-Market-Cap-Size-Value-Surges-to-Record-5-190-62-Million-by-2026-Says-Facts-Factors.html, accessed on 28 October 2022,

54 'Bitcoin Adoption Surged 880% in One Year, Emerging Markets Lead the Way', Nasdaq,https://www.nasdaq.com/articles/bitcoin-adoption-surged-880-in-one-year-emerging-markets-lead-the-way-2021-08-19, accessed on 28 October 2022

55 CoinBase, https://www.coinbase.com, accessed on 28 October 2022

56 Binance, https://www.binance.com/en, accessed on 28 October 2022

57 Bitfinex, https://www.bitfinex.com, accessed on 28 October 2022

58 'Bitcoin Energy Consumption Index', Digiconomist, https://digiconomist.net/bitcoin-energy-consumption, accessed on 28 October 2022

59 'India's Emission Capitals', Down To Earth, https://www.downtoearth.org.in/dte-infographics/61005_emission_cities_india.html, accessed on 28 October 2022,

60 'Bitcoin Causing CO2 Emissions Comparable to Hamburg', Technical University of Munich, https://www.tum.de/en/news-and-events/all-news/press-releases/details/35499, accessed on 28 October 2022

61 'On Bitcoin's Energy Consumption: A Quantitative Approach to a Subjective Question', Galaxy Digital, https://www.lopp.net/pdf/On_Bitcoin_Energy_Consumption.pdf, accessed on 28 October 2022,

62 'The Carbon Footprint of "Thank you" Emails', Statista, https://www.statista.com/chart/20189/the-carbon-footprint-of-thank-you-emails, accessed on 28 October 2022

63 'E-mail usage in the United States—Statistics and Facts', Statista, https://www.statista.com/topics/4295/e-mail-usage-in-the-united-states/#dossierKeyfigures, accessed on 28 October 2022

64 'Towards Digital Sobriety', The Shift Project, https://theshiftproject.org/wp-content/uploads/2019/03/Lean-ICT-Report_The-Shift-Project_2019.pdf, accessed on 28 October 2022

65 *How Bad are Bananas?: The Carbon Footprint of Everything*, published in 2010 by Mike Berners-Lee, Profile Books, https://howbadarebananas.com, accessed on 28 October 2022

66 'Current Trends in Sustainability of Bitcoins and Related Blockchain Technology', ResearchGate, https://www.researchgate.net/publication/321383875_Current_Trends_in_Sustainability_of_Bitcoins_and_Related_Blockchain_Technology, accessed on 28 October 2022

67 'Global E-Waste—Statistics and Facts', Statista, https://www.statista.com/topics/3409/electronic-waste-worldwide, accessed on 28 October 2022

68 'A New Circular Vision for Electronics Time for a Global Reboot', World Economic Forum, https://www3.weforum.org/docs/WEF_A_New_Circular_Vision_for_Electronics.pdf, accessed on 28 October 2022

69 'Latest Global E-Waste Statistics and What They Tell Us', The Round Up, https://theroundup.org/global-e-waste-statistics, accessed on 28 October 2022

70 'Dealing with the Discarded: E-Waste Management in India', Down To Earth, https://www.downtoearth.org.in/blog/pollution/dealing-with-the-discarded-e-waste-management-in-india-78667, accessed on 28 October 2022

71 'Electronic Waste, an Environmental Problem Exported to Developing Countries: The GOOD, the BAD and the UGLY', MDPI, https://www.mdpi.com/2071-1050/13/9/5302/htm, accessed on 28 October 2022

72 'Electronic Waste—An Emerging Threat to the Environment of Urban India', National Library of Medicine, National Center for Biotechnology Information, https://www.ncbi.nlm.nih.gov/pmc/articles/PMC3908467, accessed on 28 October 2022

73 'Apple Agrees to Pay $113 Million to Settle "Batterygate" Case Over iPhone Slowdowns', NPR, https://www.npr.org/2020/11/18/936268845/apple-agrees-to-pay-113-million-to-settle-batterygate-case-over-iphone-slowdowns, accessed on 28 October 2022

74 'Powering the Internet: Your Virtual Carbon Footprint', WebFX, https://www.webfx.com/blog/marketing/carbon-footprint-internet, accessed on 28 October 2022

75 'Why Your Internet Habits Are not as Clean as You Think', BBC, https://www.bbc.com/future/article/20200305-why-your-internet-habits-are-not-as-clean-as-you-think, accessed on 28 October 2022

76 'Is Cloud Computing Environmentally Friendly?', Business 2 Community, https://www.business2community.com/cloud-computing/is-cloud-computing-environmentally-friendly-02392578, accessed on 28 October 2022

77 'Salesforce on Track to Being the Cloud CRM Provider with the Lowest Carbon Emissions', GreenMonk, https://greenmonk.net/category/cloud, accessed on 28 October 2022

78 'Global Cloud Infrastructure Market Q4 2019 and Full Year 2019', Canalys, https://www.canalys.com/newsroom/canalys-worldwide-cloud-infrastructure-Q4-2019-and-full-year-2019, accessed on 28 October 2022

79 'The Global Tree Restoration Potential', Science, https://www.science.org/doi/10.1126/science.aax0848, accessed on 28 October 2022,

80 'Cashify', https://www.cashify.in, accessed on 28 October 2022

81 'Guidelines for Producer Responsibility Organization (PRO)', https://cpcb.nic.in/uploads/Projects/E-Waste/Guidelines_for_PRO_23.05.2018.pdf, accessed on 28 October 2022

82 'Karo Sambhav', https://www.karosambhav.com, accessed on 28 October 2022

83 'RPlanet', https://rplanet.in, accessed on 28 October 2022

84 'NAMO', https://namoewaste.com, accessed on 28 October 2022

85 'Product Environmental Report', Apple, https://www.apple.com/environment/pdf/products/iphone/iPhone_14_PER_Sept2022.pdf, accessed on 28 October 2022

Chapter 7

1 'Global Status Report for Buildings and Construction—2021', Global Alliance for Building and Construction, United Nations Environment Programme, https://globalabc.org/sites/default/files/2021-10/GABC_Buildings-GSR-2021_BOOK.pdf, accessed on 28 October 2022

2 'Renovating the Pyramid of Needs: Contemporary Extensions Built Upon Ancient Foundations', National Library of Medicine, https://www.ncbi.nlm.nih.gov/pmc/articles/PMC3161123, accessed on 28 October 2022

3 A.H. Maslow, 'A Theory of Human Motivation', *Psychological Review*, 50 (1943), pp. 370–396

4 'Global Infrastructure Construction Market—Growth, Trends, COVID-19 Impact, and Forecasts (2022 - 2027)', Mordor Intelligence, https://www.mordorintelligence.com/industry-reports/infrastructure-sector, accessed on 28 October 2022

5   'The Curious Tale of the Indian Construction Industry', 99
     Acres,   https://www.99acres.com/articles/the-curious-
     tale-of-the-indian-construction-industry.html,   accessed
     on 28 October 2022

6   'A Life Cycle Thinking Framework to Mitigate the
     Environmental Impact of Building Materials', Scientific
     Figure on ResearchGate, available from https://www.
     researchgate.net/figure/Annual-Global-Building-
     Material-Use-during-2000-2017-by-Material-and-
     Region-More-details_fig1_346372319, accessed on 28
     October 2022

7   'Department   of   Construction   Technology   and
     Management', Mizan-Tepi University, https://www.
     slideshare.net/adi5686/building-materials-and-
     environmental-impact, accessed on 28 October 2022

8   'A Life Cycle Thinking Framework to Mitigate the
     Environmental Impact of Building Materials', Cell Press
     Open Access—One Earth, https://www.cell.com/one-
     earth/pdf/S2590-3322(20)30540-6.pdf, accessed on 28
     October 2022

9   'The Cement Industry and Global Climate Change:
     Current and Potential Future Cement Industry CO2
     Emissions', Greenhouse Gas Control Technologies—
     Sixth   International   Conference,   ScienceDirect,
     https://www.sciencedirect.com/science/article/pii/
     B9780080442761501574, accessed on 28 October 2022

10   'What Is the Environmental Impact of Each Building
     Material?',   Arch   Daily,   https://www.archdaily.
     com/984663/what-is-the-environmental-impact-of-
     each-building-material, accessed on 28 October 2022

11   'Environmental Profile on Building Material Passports
     for Hot Climates', MDPI, ResearchGate, https://
     www.researchgate.net/publication/341155775_
     Environmental_Profile_on_Building_Material_
     Passports_for_Hot_Climates, accessed on 28 October
     2022

12 'Net-zero Buildings: Where Do We Stand?', WBCSD, https://www.wbcsd.org/contentwbc/download/12446/185553/1, accessed on 28 October 2022

13 'Aluminium—Not on Track', World Energy Outlook, https://www.iea.org/reports/aluminium, accessed on 28 October 2022

14 'Decarbonization Challenge for Steel', McKinsey and Company, https://www.mckinsey.com/industries/metals-and-mining/our-insights/decarbonization-challenge-for-steel, accessed on 28 October 2022

15 'Global Material Flow Analysis of Glass', *Journal of Industrial Ecology*, Wiley, https://onlinelibrary.wiley.com/doi/pdf/10.1111/jiec.13112, accessed on 28 October 2022

16 'Global Status Report for Buildings And Construction 2021', Global Alliance for Building and Construction, United Nations Environment Programme, https://globalabc.org/sites/default/files/2021-10/GABC_Buildings-GSR-2021_BOOK.pdf, accessed on 28 October 2022

17 'Global Status Report for Buildings and Construction 2021', Global Alliance for Building and Construction, United Nations Environment Programme, https://globalabc.org/sites/default/files/2021-10/GABC_Buildings-GSR-2021_BOOK.pdf, accessed on 28 October 2022

18 'Boiling Point', International Monetary Fund, https://www.imf.org/en/Publications/fandd/issues/2018/09/southeast-asia-climate-change-and-greenhouse-gas-emissions-prakash, accessed on 28 October 2022

19 'Plugging In: A Collection of Insights on Electricity Use in Indian Homes', ResearchGate, https://www.researchgate.net/publication/322791738_Plugging_In_A_Collection_of_Insights_on_Electricity_Use_in_Indian_Homes, accessed on 28 October 2022

20 'Trends in India's Residential Electricity Consumption', Center for Policy Research India, https://cprindia.org/trends-in-indias-residential-electricity-consumption, accessed on 28 October 2022

21 'We are running out of water', The World Counts, https://www.theworldcounts.com/challenges/planet-earth/freshwater/are-we-running-out-of-water, accessed on 28 October 2022

22 'Water Use and Stress', Our World in Data, https://ourworldindata.org/water-use-stress, accessed on 28 October 2022

23 Ibid

24 'What a Waste 2.0—A Global Snapshot of Solid Waste Management to 2050', The World Bank, https://datatopics.worldbank.org/what-a-waste/trends_in_solid_waste_management.html, accessed on 28 October 2022

25 'Guidelines for Municipal Solid Waste', Ministry of Housing and Urban Affairs, https://mohua.gov.in/upload/uploadfiles/files/93.pdf, accessed on 28 October 2022

26 Ibid

27 'Review on Indian Municipal Solid Waste Management Practices for Reduction of Environmental Impacts to Achieve Sustainable Development Goals', Journal of Environmental Management, Science Direct, https://www.sciencedirect.com/science/article/pii/S0301479719309405, accessed on 28 October 2022

28 'India Drowns in Construction, Demolition Debris', Down To Earth, https://www.downtoearth.org.in/news/waste/india-drowns-in-construction-demolition-debris-65110, accessed on 28 October 2022

29 'Population', United Nations, visited on 28 October 2022, https://www.un.org/en/global-issues/population

30 'Mumbai Population 2022', World Population Review, visited on 28 October 2022, https://

worldpopulationreview.com/world-cities/mumbai-population

31 'Mumbai Population 2022', World Population Review, visited on 28 October 2022, https://worldpopulationreview.com/world-cities/mumbai-population

32 'Reforms in Urban Planning Capacity in India—Final Report September 2021', NITI Aayog, https://www.niti.gov.in/sites/default/files/2021-09/UrbanPlanningCapacity-in-India-16092021.pdf, accessed on 28 October 2022

33 'Urban Populace to Increase to 38.2% by 2036', the *New Indian Express*, https://www.newindianexpress.com/nation/2020/aug/17/urban-populace-to-go-up-by-382-2184311.html, accessed on 28 October 2022

34 'Reforms in urban planning capacity in India—Final Report September 2021', NITI Aayog, https://www.niti.gov.in/sites/default/files/2021-09/UrbanPlanningCapacity-in-India-16092021.pdf, accessed on 28 October 2022

35 Ibid

36 Ibid

37 'Gujarat: Sardar Sarovar Dam, Statue of Unity Made Tadvi Adivasis Encroachers on their Own Lands', the Wire, https://thewire.in/rights/gujarat-sardar-sarovar-dam-statue-of-unity-made-tadvi-adivasis-encroachers-on-their-own-lands, accessed on 28 October 2022

38 'Blue-Green Infrastructure: An Opportunity for Indian Cities', Observer Research Foundation, https://www.orfonline.org/research/blue-green-infrastructure-an-opportunity-for-indian-cities, accessed on 28 October 2022

39 'Air Pollution, Climate Change, and Human Health in Indian Cities: A Brief Review', Frontiers, , https://www.frontiersin.org/articles/10.3389/frsc.2021.705131/full, accessed on 28 October 2022

40 'Introduction to Aerosols', Center for Aerosol Impacts on Chemistry of the Environment, https://caice.ucsd.edu/introduction-to-aerosols, accessed on 28 October 2022

41 'Air Pollution, Climate Change, and Human Health in Indian Cities: A Brief Review', Frontiers, https://www.frontiersin.org/articles/10.3389/frsc.2021.705131/full#B236, accessed on 28 October 2022

42 'Environmental Profile on Building Material Passports for Hot Climates', MDPI, ResearchGate, https://www.researchgate.net/publication/341155775_Environmental_Profile_on_Building_Material_Passports_for_Hot_Climates, accessed on 28 October 2022

43 'Nonwood Bio-based Materials—Performance of Bio-based Building Materials', ScienceDirect, https://www.sciencedirect.com/science/article/pii/B9780081009826000033?via%3Dihub, accessed on 28 October 2022

44 'Eight Sustainable Building Materials to Green Your Next Construction Project', Dumpsters, https://www.dumpsters.com/blog/green-building-materials, accessed on 28 October 2022

45 'Kochi airport becomes world's first to completely operate on solar power', Cochin International Airport Limited, https://cial.aero/pressroom/newsdetails.aspx?news_id=360, accessed on 28 October 2022

46 'A Beginner's Guide to Rainwater Harvesting', Treehugger, https://www.treehugger.com/beginners-guide-to-rainwater-harvesting-5089884, accessed on 28 October 2022

47 'Health and well-being benefits of plants', Texas A&M AgriLife Extension, https://ellisonchair.tamu.edu/health-and-well-being-benefits-of-plants, accessed on 28 October 2022

Chapter 8

1 Intergovernmental Panel on Climate Change (IPCC) Fifth Assessment Report, Working Group III,Group 10 on Industry.

2 'Perception and Deception: Human Beauty and the Brain', National Library of Medicine, National Center for Biotechnology Information, https://www.ncbi.nlm.nih.gov/pmc/articles/PMC6523404, accessed on 28 October 2022

3 'Cosmetics Market by Category, Gender, and Distribution Channel: Global Opportunity Analysis and Industry Forecast', 2021–2027, Allied Market Research, https://www.alliedmarketresearch.com/cosmetics-market, accessed on 28 October 2022

4 'Cosmetic Surgery Market', Fortune Business Insights, https://www.fortunebusinessinsights.com/cosmetic-surgery-market-102628, accessed on 28 October 2022

5 'Total number of employees of L'Oréal worldwide from 2010 to 2021', Statista, https://www.statista.com/statistics/259262/total-number-of-employees-of-loreal-worldwide, accessed on 28 October 2022

6 'Market size of the cosmetics industry across India from 2010 to 2025', Statista, https://www.statista.com/statistics/876609/india-market-size-of-cosmetics-industry | accessed on 28 October 2022

7 'Market size of beauty and personal care industry across India from 2016 to 2020, with an estimate for 2025', Statista, https://www.statista.com/statistics/1309259/india-beauty-and-personal-care-industry-market-size, accessed on 28 October 2022

8 'Lipstick Tips: How Influencers Are Making Over Beauty Marketing', Harvard Business School, https://hbswk.hbs.edu/item/lipstick-tips-how-influencers-are-making-over-beauty-marketing%20, accessed on 28 October 2022

9 'Why Social Media Marketing Will Only Become More Popular in the Beauty Industry in 2022', *Forbes*, https://www.forbes.com/sites/forbescommunicationscouncil/2022/02/15/why-social-media-marketing-will-only-become-more-popular-

in-the-beauty-industry-in-2022/?sh=28e4c59c2bb3, accessed on 28 October 2022

10 'Share of Online Beauty and Personal Care Market in India in 2021', Statista, https://www.statista.com/statistics/1309350/india-share-of-online-beauty-and-personal-care-market-by-verticals, accessed on 28 October 2022

11 'The Supply Chain Risks that Could Blemish Cosmetic Reputations', Verisk, https://www.maplecroft.com/insights/analysis/supply-chain-risks-blemish-cosmetic-reputations, accessed on 28 October 2022

12 Ibid

13 Ibid

14 'Plastic the Hidden Beauty Ingredient, The Plastic Soup Foundation', Beat the Micro Bead, https://www.beatthemicrobead.org/wp-content/uploads/2022/06/Plastic-TheHiddenBeautyIngredients.pdf, accessed on 28 October 2022

15 'World Leading Microbeads Ban Comes into Force', Department for Environment, Food and Rural Affairs and The Rt Hon Michael Gove MP, Government of UK, https://www.gov.uk/government/news/world-leading-microbeads-ban-comes-into-force, accessed on 28 October 2022

16 'Environmental Impact of Cosmetics and Beauty Products', TRVST, https://www.trvst.world/sustainable-living/environmental-impact-of-cosmetics, accessed on 28 October 2022

17 'Plastic the Hidden Beauty Ingredient, The Plastic Soup Foundation', Beat the Micro Bead, https://www.beatthemicrobead.org/wp-content/uploads/2022/06/Plastic-TheHiddenBeautyIngredients.pdf, accessed on 28 October 2022

18 'Palm Oil Products, What to Look Out For and Reasons to Avoid Palm Oil', TRVST, https://www.trvst.world/sustainable-living/eco-friendly/palm-oil-products, accessed on 28 October 2022

19 'Single-Use Plastics 101', NRDC, https://www.nrdc.org/
    stories/single-use-plastics-101, accessed on 28 October
    2022
20 'Environmental Impact of Cosmetics & Beauty Products',
    TRVST, https://www.trvst.world/sustainable-living/
    environmental-impact-of-cosmetics, accessed on 28
    October 2022
21 'Cosmetic Surgery Market', Fortune Business Insights,
    https://www.fortunebusinessinsights.com/cosmetic-
    surgery-market-102628, accessed on 28 October 2022
22 'International Survey on Aesthetic/Cosmetic Procedures
    2020', International Society of Aesthetic Plastic
    Surgery (ISAPS), https://www.isaps.org/wp-content/
    uploads/2022/01/ISAPS-Global-Survey_2020.pdf,
    accessed on 28 October 2022
23 'International Survey on Aesthetic/Cosmetic Procedures
    2020', International Society of Aesthetic Plastic
    Surgery (ISAPS), https://www.isaps.org/wp-content/
    uploads/2022/01/ISAPS-Global-Survey_2020.pdf,
    accessed on 28 October 2022
24 'Social Media and its Effects on Beauty', Mavis
    Henriques and Debasis Patnaik, IntechOpen, https://
    www.intechopen.com/chapters/73271, accessed on 28
    October 2022
25 'Eating Disorder Statistics 2022', The Checkup by
    SingleCare, https://www.singlecare.com/blog/news/
    eating-disorder-statistics, accessed on 28 October
    2022
26 'Average Annual Spend on Cosmetic Products
    Among Female Consumers in Selected Countries
    Worldwide in 2020', Statista, https://www.statista.
    com/statistics/1224276/average-annual-spend-on-
    cosmetics-among-women-worldwide, accessed on 28
    October 2022
27 'Beauty and Personal Care Market in India, Ministry of
    Economy and Industry Foreign Trade Administration',
    https://www.export.gov.il/files/cosmetics/

ReportonIndianCosmeticsIndustry.pdf, accessed on 28 October 2022

28 'Environmental Impact of Cosmetics & Beauty Products', TRVST, https://www.trvst.world/sustainable-living/ environmental-impact-of-cosmetics, accessed on 28 October 2022

29 Ibid

30 Ibid

31 'International Survey on Aesthetic/Cosmetic Procedures 2020', International Society of Aesthetic Plastic Surgery (ISAPS), https://www.isaps.org/wp-content/ uploads/2022/01/ISAPS-Global-Survey_2020.pdf, accessed on 28 October 2022

32 'Environmental Impact of Cosmetics & Beauty Products', TRVST, https://www.trvst.world/sustainable-living/ environmental-impact-of-cosmetics, accessed on 28 October 2022

33 'Child Labor and Forced Labor Reports', Bureau of International Labor Affairs, https://www.dol.gov/ agencies/ilab/resources/reports/child-labor/cote-divoire, accessed on 28 October 2022

34 'The Truth About How Ethical Your Beauty Products Really Are', Refinery29, https://www.refinery29.com/en-gb/2018/06/202718/natural-beauty-products-ethical-risks, accessed on 28 October 2022

35 'Environmental Impact of Cosmetics and Beauty Products', TRVST, https://www.trvst.world/sustainable-living/environmental-impact-of-cosmetics, accessed on 28 October 2022

36 'Arguments Against Animal Testing', Cruelty Free International, https://crueltyfreeinternational.org/ about-animal-testing/arguments-against-animal-testing, accessed on 28 October 2022

37 'Environmental Impact of Cosmetics and Beauty Products', TRVST, https://www.trvst.world/sustainable-living/environmental-impact-of-cosmetics, accessed on 28 October 2022

38 'Arguments Against Animal Testing', Cruelty Free International, https://crueltyfreeinternational.org/about-animal-testing/arguments-against-animal-testing, accessed on 28 October 2022

39 'Environmental Impact of Cosmetics and Beauty Products', TRVST, https://www.trvst.world/sustainable-living/environmental-impact-of-cosmetics, accessed on 28 October 2022

40 'Peta India Applauds New Rules for Strengthening Import Ban on Animal-Tested Cosmetics', India Today, Peta, https://www.indiatoday.in/india/story/peta-india-applauds-new-rules-for-strengthening-import-ban-on-animal-tested-cosmetics-1751661-2020-12-21, accessed on 28 October 2022

41 'A Step Forward on Sustainability in the Cosmetics Industry: A Review', *Journal of Cleaner Production*, https://www.sciencedirect.com/science/article/abs/pii/S0959652619309655, accessed on 28 October 2022

42 'Demand for "Natural or Organic Ingredients" or "Sustainable" Beauty Products Worldwide as of September 2019', Statista, https://www.statista.com/statistics/803595/global-demand-for-natural-organic-environmental-friendly-cosmetics, accessed on 28 October 2022

Chapter 9

1 Intergovernmental Panel on Climate Change (IPCC) Fifth Assessment Report, Working Group III, Chapter 10 on Industry.

2 'Plastic Leakage and Greenhouse Gas Emissions Are Increasing', Organisation for Economic Co-operation and Development, https://www.oecd.org/environment/plastics/increased-plastic-leakage-and-greenhouse-gas-emissions.htm, accessed on 28 October 2022

3 'Annual Production of Plastics Worldwide from 1950 to 2020', Statista, https://www.statista.com/

statistics/282732/global-production-of-plastics-since-1950, accessed on 28 October 2022

4   'Cumulative Global Production of Plastics', Our World in Data, https://ourworldindata.org/grapher/cumulative-global-plastics, accessed on 28 October 2022

5   'Annual Production of Plastics Worldwide from 1950 to 2020', Statista, https://www.statista.com/statistics/282732/global-production-of-plastics-since-1950, accessed on 28 October 2022

6   'Plastic Pollution', Our World in Data, https://ourworldindata.org/plastic-pollution, accessed on 28 October 2022

7   'Plastic Pollution', Our World in Data, https://ourworldindata.org/plastic-pollution#cumulative-production, accessed on 28 October 2022

8   'Ten Tons of Trash Removed from Mount Everest', Breaking Asia, https://www.breakingasia.com/news/ten-tons-of-trash-removed-from-mount-everest, accessed on 28 October 2022

9   'The Biogeography of the Plastisphere: Implications for Policy', John Wiley & Sons, Inc,. https://esajournals.onlinelibrary.wiley.com/doi/abs/10.1890/150017, accessed on 28 October 2022

10 'The Future's not in Plastics', Carbon Tracker, https://zerokonferansen.no/wp-content/uploads/2020/12/The_Futures_Not_in_Plastics-Carbon-Tracker.pdf, accessed on 28 October 2022

11 'The Future's Not in Plastics', Carbon Tracker, https://zerokonferansen.no/wp-content/uploads/2020/12/The_Futures_Not_in_Plastics-Carbon-Tracker.pdf, accessed on 28 October 2022

12 'Making Plastics in India: Trends in the Industry', Centre for Financial Accountability (CFA), https://www.cenfa.org/making-plastics-in-india-trends-in-the-industry, accessed on 28 October 2022

13 'Making Plastics in India: Trends in the Industry', Centre for Financial Accountability (CFA), https://www.cenfa.

org/making-plastics-in-india-trends-in-the-industry, accessed on 28 October 2022

14 'The Problem of Plastics', Centre for Science and Environment, https://cdn.cseindia.org/attachments/0.57139300_1570431848_Factsheet1.pdf | accessed on 28 Oct 2022

15 'Plastic waste generation by industrial sector, 2015', Our World in Data, https://ourworldindata.org/grapher/plastic-waste-by-sector, accessed on 28 October 2022

16 'Primary Plastic Production by Industrial Sector, 2015', Our World in Data, https://ourworldindata.org/grapher/plastic-production-by-sector, accessed on 28 October 2022

17 'Gems and Jewellery Industry and Products', India Brand Equity Foundation, https://www.ibef.org/exports/plastic-industry-india%20-%20India%20Brand%20Equity%20Foundation%20-%20https://www.ibef.org/exports/plastic-industry-india%20%7C%20accessed%20on%2028%20Oct%202022, accessed on 28 October 2022

18 'An Indian Consumes 11kg Plastic Every Year and an Average American 109kg', Down To Earth, https://www.downtoearth.org.in/news/waste/an-indian-consumes-11-kg-plastic-every-year-and-an-average-american-109-kg-60745, accessed on 28 October 2022

19 'Gems and Jewellery Industry and Products', India Brand Equity Foundation, https://www.ibef.org/exports/plastic-industry-india%20-%20India%20Brand%20Equity%20Foundation%20-%20https://www.ibef.org/exports/plastic-industry-india%20%7C%20visited%20on%2028%20Oct%202022, accessed on 28 October 2022

20 'Plastic Waste Management Rules', Central Pollution Control Board, Ministry of Environment, Forest and Climate Change, Government of India, https://cpcb.nic.in/uploads/plasticwaste/Annual_Report_2018-19_PWM.pdf, accessed on 28 October 2022

21 'India is Generating Much More Plastic Waste Than It Reports', BQ Prime, *Bloomberg,* https://www.bqprime.com/global-economics/india-is-generating-much-more-plastic-waste-than-it-reports-heres-why, accessed on 28 October 2022

22 'Ban on Identified Single Use Plastic Items from 1st July 2022', Ministry of Environment, Forest and Climate Change, https://pib.gov.in/PressReleaseIframePage.aspx?PRID=1837518, accessed on 28 October 2022

23 'Single-use Plastic Ban: Reading the Fine Print Reveals Ominous Loopholes', Down To Earth, https://www.downtoearth.org.in/blog/waste/single-use-plastic-ban-reading-the-fine-print-reveals-ominous-loopholes-78496, accessed on 28 October 2022

24 'The Problem of Plastics', Centre for Science and Environment, https://cdn.cseindia.org/attachments/0.57139300_1570431848_Factsheet1.pdf, accessed on 28 October 2022

25 'Number of Landfills in India in Financial Year 2020, by State', Statista, https://www.statista.com/statistics/1168458/india-number-of-landfills-by-state, accessed on 28 October 2022

26 'Plastic Recycling Codes Explained, Types of Plastic and the Applications of Recycled Plastics', AZO Materials, https://www.azom.com/article.aspx?ArticleID=4425, accessed on 28 October 2022

27 'HDPE Can Be Recycled at Least 10 Times', Microdyne Plastics Inc, http://microdyneplastics.com/2018/02/hdpe-can-recycled-least-10-times, accessed on 28 October 2022

28 'Which Plastic Can Be Recycled?' Plastics for Change, https://www.plasticsforchange.org/blog/which-plastic-can-be-recycled, accessed on 28 October 2022

29 'Our Planet Is Choking on Plastic', United Nations Environment Programme, https://www.unep.org/interactives/beat-plastic-pollution, accessed on 28 October 2022

30 'Plastic Pollution, Our World in Data', https://ourworldindata.org/plastic-pollution#plastic-disposal-methods, accessed on 28 October 2022

31 'Plastic and Climate: The Hidden Costs of a Plastic Planet', Plastic Pollution Coalition, https://www.ciel.org/wp-content/uploads/2019/05/Plastic-and-Climate-FINAL-2019.pdf, accessed on 28 October 2022

32 'FAQs on Plastics', Our World in Data, https://ourworldindata.org/faq-on-plastics#are-plastic-alternatives-better-for-the-environment, accessed on 28 October 2022

33 'Plastic Pollution', Our World in Data, https://ourworldindata.org/plastic-pollution#microplastics-impacts-on-health, accessed on 28 October 2022

34 'Plastic Pollution', Our World in Data, https://ourworldindata.org/plastic-pollution#microplastics-impacts-on-health, accessed on 28 October 2022

35 'Plastic Pollution', Our World in Data, https://ourworldindata.org/plastic-pollution#microplastics-impacts-on-health, accessed on 28 October 2022

36 'From Pollution to Solution', United Nations Environment Programme, https://wedocs.unep.org/bitstream/handle/20.500.11822/36963/POLSOL.pdf, accessed on 28 October 2022

37 'Adoption, Use and Environmental Impact of Feminine Hygiene Products Among College Going Girls of Udaipur', Research Gate, https://www.researchgate.net/publication/327842879_Adoption_Use_and_Environmental_Impact_of_Feminine_Hygiene_Products_among_College_Going_Girls_of_Udaipur, accessed on 28 October 2022

38 'A State-of-the-Art Review of Indigenous Peoples and Environmental Pollution', *Integrated Environmental Assessment and Management* Volume 16, Number 3, pp. 324–341, ResearchGate, https://www.researchgate.net/publication/338097003_A_State-of-the-Art_Review_

of_Indigenous_Peoples_and_Environmental_Pollution, accessed on 28 October 2022

39 'Our Planet Is Choking on Plastic', United Nations Environment Programme, https://www.unep.org/interactives/beat-plastic-pollution, accessed on 28 October 2022

40 'Plastic Suffocation: Climate Change Threatens Indigenous Populations and Traditional Ecological Knowledge', Samantha Chisholm Hateld, Oregon Climate Change Research Institute, Oregon State University, USA, https://jmic.online/issues/v8n2/pdf/Journal%20of%20Marine%20and%20Island%20Cultures%20v8n2-1%20-%20Chisholm%20Hatfield.pdf, accessed on 28 October 2022

41 'COVID Pollution: Impact of COVID-19 Pandemic on Global Plastic Waste Footprint', Science Direct, https://www.sciencedirect.com/science/article/pii/S2405844021004485, accessed on 28 October 2022

42 'Marine Plastic Pollution Has Increased Tenfold Since 1980, says Biodiversity Report', Mongabay, https://india.mongabay.com/2019/05/marine-plastic-pollution-has-increased-tenfold-since-1980-says-biodiversity-report, accessed on 28 October 2022

43 'Tonnes of COVID-19 Healthcare Waste Expose Urgent Need to Improve Waste Management Systems', World Health Organization, https://www.who.int/news/item/01-02-2022-tonnes-of-covid-19-health-care-waste-expose-urgent-need-to-improve-waste-management-systems, accessed on 28 October 2022

44 'COVID Pollution: Impact of COVID-19 pandemic on Global Plastic Waste Footprint', Science Direct, https://www.sciencedirect.com/science/article/pii/S2405844021004485, accessed on 28 October2022

45 'Plastic Waste Release Caused by COVID-19 and its Fate in the Global Ocean', National Library of Medicine, https://pubmed.ncbi.nlm.nih.gov/34751160, accessed on 28 October 2022

46 'Fighting for Trash Free Seas', Ocean Conservancy, https://oceanconservancy.org/trash-free-seas, accessed on 28 October 2022

47 'Plastic Pollution', Our World in Data, https://ourworldindata.org/plastic-pollution#microplastics-impacts-on-health, accessed on 28 October 2022

48 'Plastics in the Marine Environment', Research Gate, https://www.researchgate.net/publication/308039665_Plastics_in_the_Marine_Environment, accessed on 28 October 2022

49 'Plastic Pollution', Our World in Data, https://ourworldindata.org/plastic-pollution, accessed on 28 October 2022

50 'Probability of Mismanaged Plastic Waste Being Emitted to Ocean 2019', Our World in Data, https://ourworldindata.org/grapher/probability-mismanaged-plastic-ocean, accessed on 28 October 2022

51 'Plastic Pollution', Our World in Data, https://ourworldindata.org/plastic-pollution, accessed on 28 October 2022

52 'Plastic Pollution', Our World in Data, https://ourworldindata.org/plastic-pollution, accessed on 28 Oct 2022

53 'Plastic Pollution', Our World in Data, https://ourworldindata.org/plastic-pollution#the-great-pacific-garbage-patch-gpgp, accessed on 28 October 2022

54 '7 Eye-opening Facts About the Pacific Garbage Patch and the 5 Gyres', Healthy Human, https://healthyhumanlife.com/blogs/news/pacific-garbage-patch-facts, accessed on 28 October 2022

55 'Great Pacific Garbage Patch', National Geographic, https://education.nationalgeographic.org/resource/great-pacific-garbage-patch, accessed on 28 October 2022

56 'How Can We Destroy the Great Pacific Garbage Patch?', WWF, https://www.wwf.org.au/news/blogs/how-can-we-destroy-the-great-pacific-garbage-patch, accessed on 28 October 2022

57 'Plastic Pollution', Our World in Data, https://ourworldindata.org/plastic-pollution#the-great-pacific-garbage-patch-gpgp, accessed on 28 October 2022

58 'Plastic Pollution', Our World in Data, https://ourworldindata.org/plastic-pollution#microplastics-impacts-on-health, accessedon 28 October 2022

59 'Reduce Your Plastic Waste', Her Planet Earth, https://www.herplanetearth.com/reduce-plastic-waste.html, accessed on 28 October 2022

60 'Plastic Waste Collected for Recycling', Organization for Economic Co-operation and Development, https://stats.oecd.org/viewhtml.aspx?datasetcode=PLASTIC_WASTE_6&lang=en, accessed on 28 October 2022

61 'The Dirty Truth About Floss', Tree Bird, https://treebirdeco.com/blogs/blog/the-dirty-truth-about-floss, accessed on 28 October 2022

Chapter 10

1 'Data Commons', Place Explorer, India, Overview, https://datacommons.org/place/country/IND, accessed on 28 October 2022

2 'Census of India 2011', Language Atlas, INDIA, https://censusindia.gov.in/nada/index.php/catalog/42561/download/46187/Language_Atlas_2011.pdf, accessed on 28 October 2022

3 'Census of India 2011', Language Atlas, INDIA, https://censusindia.gov.in/nada/index.php/catalog/42561/download/46187/Language_Atlas_2011.pdf, accessed on 28 October 2022

4 'Thanksgiving Dinner's Carbon Footprint', *Chicago Tribune*, https://www.chicagotribune.com/lifestyles/health/sns-green-thanksgiving-carbon-footprint-story.html | accessed on 28 October 2022

5 'Re-imagining Holiday Habits for a More Sustainable Future', Impakter, https://impakter.com/reimagining-

holiday-habits-for-a-more-sustainable-future,   accessed on 28 October 2022

6   'Black Friday 2022 India', Black Friday Global, https://black-friday.global/en-in, accessed on 28 October 2022

7   'Black Friday 2022 India', Black Friday Global, https://black-friday.global/en-in, accessed on 28 October 2022

8   'Black Friday Death Count', http://blackfridaydeathcount.com, accessed on 28 October 2022

9   'Association Between Changes in Air Quality and Hospital Admissions During the Holi Festival', Springer Nature Switzerland AG 2019, https://link.springer.com/article/10.1007/s42452-019-0165-5, accessed on 28 October 2022

10   'Create Memories, not Trash: Why Green Weddings Are the Need of the Hour', the *Indian Express*, https://indianexpress.com/article/lifestyle/life-style/create-memories-not-trash-green-weddings-eco-friendly-sustainable-waste-management-plastics-carboon-footprints-7929511, accessed on 28 October 2022

11   'Household Social Consumption on Education in India', Ministry of Statistics and Programme Implementation, Government of India, http://164.100.161.63/sites/default/files/publication_reports/Report_585_75th_round_Education_final_1507_0.pdf, accessed on 28 October 2022

12   'Number of Enrolled Students in India as of 2018, by School Type', Statista, https://www.statista.com/statistics/1175285/india-number-of-enrolled-students-by-school-type, accessed on 28 October 2022

13   'India's Education Budget Cannot Fund Proposed New Education Policy', IndiaSpend, https://www.indiaspend.com/indias-education-budget-cannot-fund-proposed-new-education-policy, accessed on 28 October 2022

14   'Teacher Salaries in India', Glassdoor, https://www.glassdoor.co.in/Salaries/teacher-salary-SRCH_KO0,7.htm, accessed on 28 October 2022

15   'Indians are Spending Enormously on Education Even with Few Jobs in Sight', Business Insider India, https://

www.businessinsider.in/education/news/average-education-expenditure-in-india-increases-fourfold-to-8331-per-student/articleshow/72282009.cms, accessed on 28 October 2022

16   'Indians are Spending Enormously on Education Even with Few Jobs in Sight', Business Insider India, https://www.businessinsider.in/education/news/average-education-expenditure-in-india-increases-fourfold-to-8331-per-student/articleshow/72282009.cms, accessed on 28 October 2022

17   'Census of India 2011, Language Atlas', INDIA, https://censusindia.gov.in/nada/index.php/catalog/42561/download/46187/Language_Atlas_2011.pdf,   accessed on 28 October 2022

18   'Indians are Spending Enormously on Education Even with Few Jobs in Sight', Business Insider India, https://www.businessinsider.in/education/news/average-education-expenditure-in-india-increases-fourfold-to-8331-per-student/articleshow/72282009.cms, accessed on 28 October 2022

19   'Household Social Consumption on Education in India', Ministry of Statistics and Programme Implementation, Government of India, http://164.100.161.63/sites/default/files/publication_reports/Report_585_75th_round_Education_final_1507_0.pdf, accessed on 28 October 2022

20   'Household Social Consumption on Education in India', Ministry of Statistics and Programme, Implementation, Government of India, http://164.100.161.63/sites/default/files/publication_reports/Report_585_75th_round_Education_final_1507_0.pdf, accessed on 28 October 2022

21   'Environmental Education in Schools: The Indian Scenario', Research Gate, https://www.researchgate.net/publication/255614279_Environmental_Education_in_Schools_The_Indian_Scenario,   accessed   on   28 October 2022

22 'Environment Education in Indian, Challenges and Opportunities', Government of India, Ministry of Environment, Forest and Climate Change, India, https://casopis.hrcpo.com/volume-11-issue-4-puri-et-al, accessed on 28 October 2022

23 'Despite Global Concerns About Democracy, More than Half of Countries Are Democratic', Pew Research Center, https://www.pewresearch.org/fact-tank/2019/05/14/more-than-half-of-countries-are-democratic, accessed on 28 October 2022

24 'Many Across the Globe are Dissatisfied with How Democracy Is Working', Pew Research Center, https://www.pewresearch.org/global/2019/04/29/many-across-the-globe-are-dissatisfied-with-how-democracy-is-working, accessed on 28 October 2022

25 'Many Across the Globe are Dissatisfied with How Democracy is Working', Pew Research Center, https://www.pewresearch.org/global/2019/04/29/many-across-the-globe-are-dissatisfied-with-how-democracy-is-working, accessed on 28 October 2022

26 'Funding Elections in India: Whose Money has the Most Influence?' *Hindustan Times*, https://www.hindustantimes.com/india-news/funding-elections-in-india-whose-money-has-the-most-influence/story-KlBvlvtVhdfAPy6hRes3PJ.html, accessed on 28 October 2022

27 'Green Employment Opportunities in Emerging Markets* Across Key Sectors between 2020 and 2030', Statista, https://www.statista.com/statistics/1258751/new-direct-jobs-by-sector, accessed on 28 October 2022

28 'Companies Who Report on Sustainability Worldwide from 1993 to 2020', Statista, https://www.statista.com/statistics/1232295/global-sustainability-reporting-growth-rate, accessed on 28 October 2022

29 'World Population Growth', Our World in Data, https://ourworldindata.org/world-population-growth, accessed on 28 October 2022

30 'The Best Way to Reduce Your Carbon Footprint Is
    One the Government Isn't Telling You About', Science,
    https://www.science.org/content/article/best-way-
    reduce-your-carbon-footprint-one-government-isn-t-
    telling-you-about, accessed on 28 October 2022
31 'The Best Way to Reduce Your Carbon Footprint is
    One the Government Isn't Telling You About', Science,
    https://www.science.org/content/article/best-way-
    reduce-your-carbon-footprint-one-government-isn-t-
    telling-you-about, accessed on 28 October 2022

Glossary

Terms to Know

A:

Aerosols: A suspension of airborne solid such as dust or liquid particles such as mist droplets (of size between a few nanometres and 10μm) in the air. (IPCC)

Aerosols simplified: An aerosol is a suspension of fine solid particles or liquid droplets in air or another gas. Aerosols can be natural or anthropogenic.

Agrarian: Anything that is of or relating to agricultural fields or lands. (Webster)

Agrarian simplified: Relating to cultivated land or the cultivation of land.

Air pollution: Degradation of air quality with negative effects on human health or the natural or built environment due to the introduction of substances (gases, aerosols) into the atmosphere by natural processes or human activity, which have a direct (primary pollutants) or indirect (secondary pollutants) harmful effect. (IPCC)

Alternative fuels: a fuel for internal combustion engines that is derived partly or wholly from a source other than petroleum and that is less damaging to the environment than traditional fuels. (Webster).

Anaerobic: living, active, occurring, or existing in the absence of free oxygen. (Webster)

Anthropogenic: Anything that is a result of or produced by human activities. (IPCC)

Antifouling system: Antifouling systems can be defined as the coating, paint, and surface treatment used on a solid (e.g., ship hull) to control or prevent the attachment of unwanted organisms. (Transport Stylersen)

Atmosphere: The gaseous envelope surrounding the earth. (IPCC)

Atmosphere simplified: An atmosphere is the layers of gases surrounding a planet or other celestial body. Earth's atmosphere is composed of about 78 per cent nitrogen, 21 per cent oxygen, and 1 per cent other gases

Aquaculture: Aquaculture is like farming, except that it takes place in the water. It refers to the breeding, raising, and harvesting of fish, shellfish, and aquatic plants.

Aquifers: An aquifer is a body of rock and sediment that holds groundwater.

B:

Biocapacity: The capacity of ecosystems to produce useful biological materials and to absorb waste materials generated by humans, using current management schemes and extraction technologies. (WWF)

Biodegradable: Anything capable of being broken down especially into smaller particles by microorganisms. (Webster)

Biodegradable simplified: Any substance or object that is capable of being decomposed naturally by bacteria or other living organisms and thereby avoiding pollution.

Biodiversity: The variability among living organisms from all sources including, inter alia, terrestrial, marine and other aquatic ecosystems and the ecological complexes of which they are part; this includes diversity within species, between species and of ecosystems. (CBD)

Biodiversity simplified: Biodiversity is the variety of all life forms on Earth—the different plants, animals and micro-organisms, their genes and the terrestrial, marine and freshwater ecosystems of which they are a part.

Biofuel: A fuel, generally in liquid form, produced from biomass. (IPCC)

Biomass: The weight or total quantity of living organisms of one animal or plant species (species biomass) or of all the species in a community (community biomass), commonly referred to a unit area or volume of habitat. (Encyclopedia Britannica)

Biomass simplified: Biomass is organic, meaning it is made of material that comes from living organisms, such as plants and animals.

Blockchain: A blockchain is a digitally distributed, decentralized, public ledger that exists across a network. A blockchain is essentially a digital ledger of transactions that is duplicated and distributed across the entire network of computer systems on the blockchain.

C:

Capitalism: An economic system characterized by private or corporate ownership of capital goods, by investments that are determined by private decision and by prices, production and the distribution of goods that are determined mainly by competition in a free market. (Webster)

Carbon budget: This term refers to three concepts in the literature: (1) an assessment of carbon cycle sources and sinks on a global level, through the synthesis of evidence for fossil fuel and cement emissions, land-use change emissions, ocean and land CO_2 sinks, and the resulting atmospheric CO_2 growth rate. (IPCC)

Carbon budget simplified: A tolerable quantity of greenhouse gas emissions that can be emitted in total over a specified time.

Carbon colonialism: The ability of wealthier countries to effectively outsource emissions to less wealthy ones has been described as 'carbon colonialism'.

Carbon colonialism simplified: It is the practice of developed countries shifting the burden of reducing carbon emissions to developing countries.

Carbon dioxide (CO_2): A naturally occurring gas, CO_2 is also a by-product of burning fossil fuels (such as oil, gas and coal), of burning biomass, of land-use changes (LUC) and of industrial processes (e.g., cement production). It is the principal anthropogenic greenhouse gas (GHG) that affects the Earth's radiative balance. (IPCC). Also when we breathe out, we release carbon dioxide into the air which is then used by the plants to make their food and the cycle starts all over again.

Carbon footprint: The total amount of greenhouse gases that are emitted into the atmosphere each year by a person, family, building, organization or company. A person's carbon footprint includes greenhouse gas emissions from fuel that an individual burns directly, such as by heating a home or riding in a car. It also includes greenhouse gases that come from producing the goods or services that the individual uses, including emissions from power plants that make electricity, factories that make products and landfills where trash gets sent. (Environment Statistics Glossary, Ministry of Statistics and Programme Implementation, GoI)

Carbon offsetting: Any activity that compensates for the emission of carbon dioxide (CO_2) or other greenhouse gases (measured in carbon dioxide equivalents [CO_2e]) by providing for an emission reduction elsewhere. (Encyclopedia Britannica)

Carbon profile: The measurement of overall amount of CO_2 and other GHG emissions. (IPCC)

Carbon sequestration: The process of storing carbon in a carbon Pool. (IPCC)

Carbon sinks: A reservoir or pool (natural or human, for example: in soil, ocean, and plants) where a greenhouse gas, an aerosol or a precursor of a greenhouse gas is stored. (IPCC)

Carbon stock: The quantity of carbon contained in a 'pool', meaning a reservoir or system which has the capacity to accumulate or release carbon. (Greenfacts)

Cardiovascular: Anything of, relating to, or involving the heart and blood vessels. (Webster)

Climate change: Climate change refers to a change in the state of the climate that can be identified (e.g., by using statistical tests) by changes in the mean and/or the variability of its properties and that persists for an extended period, typically decades or longer. Climate change may be due to natural internal processes or external forcings such as modulations of the solar cycles, volcanic eruptions and persistent anthropogenic changes in the composition of the atmosphere or in land use. (IPCC)

Climate: Climate in a narrow sense is usually defined as the average weather, or more rigorously, as the statistical description in terms of the mean and variability of relevant quantities over a period of time ranging from months to thousands or millions of years. (IPCC)

CO_2 equivalent (CO_2-eq): The amount of carbon dioxide (CO_2) emission that would cause the same integrated radiative forcing or temperature change, over a given time horizon, as an emitted amount of a greenhouse gas (GHG) or a mixture of GHGs. (IPCC)

CO_2 equivalent simplified: A metric measure used to compare the emissions from various greenhouse gases on the basis of their global-warming potential (GWP), by converting amounts of other gases to the equivalent amount of carbon dioxide with the same global warming potential.

Combustion: A usually rapid chemical process (such as oxidation) that produces heat and usually light in the form of flame. (Webster)

Composting: Composting is the natural process of recycling organic matter, such as leaves and food scraps, into a valuable fertilizer that can enrich soil and plants. (NRDC)

Compound annual growth rate (CAGR): CAGR is the annualized average rate of growth between two given years, assuming growth takes place at an exponentially compounded rate. The CAGR between given years X and Z, where Z - X = N, is the number of years between the two given years, is calculated as follows:

CAGR, year X to year Z = [(value in year Z/value in year X) ^ (1/N)-1]

Conservation: Study of the loss of Earth's biological diversity and the ways this loss can be prevented. (Encyclopedia Britannica)

Consumption: The act or process of consuming. (Webster)

Contamination: The presence of a constituent, impurity, or some other undesirable element that spoils, corrupts, infects, makes unfit, or makes inferior a natural environment. (Webster)

Cryptocurrencies: Any form of currency that only exists digitally, usually has no central issuing or regulating authority but instead uses a decentralized system to record transactions and manage the issuance of new units and relies on cryptography to prevent counterfeiting and fraudulent transactions. (Webster)

D:

Decarbonization: The process by which countries, individuals or other entities aim to achieve zero fossil carbon existence. Typically refers to a reduction of the carbon emissions associated with electricity, industry and transport. (IPCC)

Deforestation: Conversion of forest to non-forest. (IPCC)

Downcycling: to recycle (something) in such a way that the resulting product is of a lower value than the original item: to create an object of lesser value from (a discarded object of higher value). (Webster)

E:

Ecological footprint: Measure of the demands made by a person or group of people on global natural resources. (Encyclopedia Britannica)

Ecology: A branch of science concerned with the interrelationship of organisms and their environments. (Encyclopedia Britannica)

Ecosphere: The parts of the universe habitable by living organisms. (Webster)

Eco-tourism: The practice of touring natural habitats in a manner meant to minimize ecological impact. (Webster)

Electric vehicle (EV): A vehicle whose propulsion is powered fully or mostly by electricity. (IPCC)

Endangered: Being or relating to an endangered species. (Webster)

Energy efficiency: The ratio of output or useful energy or energy services or other useful physical outputs obtained from a system, conversion process, transmission or storage activity to the input of energy. (IPCC)

Environmental science: An interdisciplinary academic field that draws on ecology, geology, meteorology, biology, chemistry, engineering, and physics to study environmental problems and human impacts on the environment. (Encyclopedia Britannica)

ESG risk: An environmental, social, or governance event, or condition that, if it occurs, could cause an actual or a potential material negative impact on the value of the investment arising from an adverse sustainability impact. (Deloitte)

Ethernet: A computer network architecture consisting of various specified local-area network protocols, devices, and connection methods. (Webster)

Eutrophication: The process by which a body of water becomes enriched in dissolved nutrients (such as phosphates) that stimulate the growth of aquatic plant life, usually resulting in the depletion of dissolved oxygen. (Webster)

E-waste: Waste consisting of discarded electronic products (such as computers, televisions, and cell phones). (Webster)

Exploitation: An act or instance of exploiting (Webster). In this case exploiting natural resources.

Extended producer responsibility: The commitment made by a producer to facilitate a reverse collection mechanism and recycling of end of life, post-consumer waste. The objective is to circle it back into the system to recover resources embedded in the waste. (Saahas)

F:

Fin erosion: Fin erosion or fin rot is a disease in fish which occurs because of a bacterial or fungal infections. (NOAA)

Food miles: The distance traveled by any food before it reaches your plate. (DTE)

Food security: A situation that exists when all people, at all times, have physical, social and economic access to sufficient, safe and nutritious food that meets their dietary needs and food preferences for an active and healthy life. (IPCC)

Food wastage: Food wastage encompasses food loss (the loss of food during production and transportation) and food waste (the waste of food by the consumer). (IPCC)

Forest: A vegetation type dominated by trees. (IPCC)

Fossil fuels: Any of a class of hydrocarbon-containing materials of biological origin occurring within Earth's crust that can be used as a source of energy. (Encyclopedia Britannica)

Free market: An unregulated system of economic exchange, in which taxes, quality controls, quotas, tariffs, and other forms of centralized economic interventions by government either do not exist or are minimal. (Encyclopedia Britannica)

G:

Gentrification: A process in which a poor area (as of a city) experiences an influx of middle-class or wealthy people who renovate and rebuild homes and businesses and which often results in an increase in property values and the displacement of earlier, usually poorer residents. (Webster)

Global warming: The estimated increase in global mean surface temperature (GMST) averaged over a 30-year period, or the 30-year period centered on a particular year or decade, expressed relative to pre-industrial levels unless otherwise specified. (IPCC)

Global-Warming Potential (GWP): Developed to allow comparisons of the global warming impacts of different gases. Specifically, it is a measure of how much energy the emissions of 1 ton of a gas will absorb over a given period of time, relative to the emissions of 1 ton of carbon dioxide (CO_2).

Globalisation: The development of an increasingly integrated global economy marked especially by free trade, free flow of capital, and the tapping of cheaper foreign labour markets. (Webster).

Green stocks: Green stocks are companies that invest in environmentally friendly alternative-energy sources.

Greenhouse effect: Trapping and build–up of heat in the atmosphere (troposphere) near the Earth's surface. (Environment Statistics Glossary, Ministry of Statistics and Programme Implementation, GoI)

Greenhouse gas (GHG): Greenhouse gases are those gaseous constituents of the atmosphere, both natural and anthropogenic, that absorb and emit radiation at specific wavelengths within the spectrum of terrestrial radiation emitted by the Earth's surface, the atmosphere itself and by clouds. (IPCC)

Greenwashing: A form of deceptive marketing in which a company, product, or business practice is falsely or excessively promoted as being environmentally friendly. (Encyclopedia Britannica)

Gross domestic product (GDP): The sum of gross value added, at purchasers' prices, by all resident and non-resident producers in the economy, plus any taxes and minus any subsidies not included in the value of the products in a country or a geographic region for a given period, normally one year. (IPCC)

H:

Heavy metals: A metal of high specific gravity. (Webster)

Human-animal conflict: It is defined as the struggles that arise from people and animals coming into contact—often leads to people killing animals in self-defence, or as pre-emptive or retaliatory killings, which can drive species to extinction. (UNEP)

I:

Internal combustion engine (ICE): Any of a group of devices (in an engine) in which the reactants of combustion (oxidizer and fuel) and the products of combustion serve as the working fluids of the engine. (Encyclopedia Britannica)

Indigenous: The original inhabitants of a particular geographic region. (Encyclopedia Britannica)

Internet of Things (IoT): The networking capability that allows information to be sent to and received from objects and devices (such as fixtures and kitchen appliances) using the Internet. (Encyclopedia Britannica)

L:

Land clearing: Land Clearing is removing trees, stumps, and other vegetation from wooded areas. (Collins Dictionary)

Land use: Land use refers to all arrangements, activities and inputs undertaken in a certain land cover type. (IPCC)

Landfill: A system of trash and garbage disposal in which the waste is buried between layers of earth to build up low-lying land. (Webster)

Land-use change (LUC): Land-use change involves a change from one land use category to another. (IPCC)

Leaching: To draw out or remove as if by percolation. (Webster)

Livelihood: The resources used and the activities undertaken in order to live. (IPCC)

M:

Macroplastic: Relatively large particles of plastic found especially in the marine environment (typically more than about 5 mm). (Encyclopedia Britannica)

Marginalisation: To relegate to an unimportant or powerless position within a society or group. (Webster)

Maximalism: The opposite of minimalism: a tendency toward excess (Webster), in this case excess usage of resources.

Methane (CH_4): One of the six greenhouse gases (GHGs) to be mitigated under the Kyoto Protocol and is the major component of natural gas and associated with all hydrocarbon fuels. (IPCC)

Micro mobility vehicles (MMVs): Micromobility vehicles refers to short-distance transport, usually less than 5 miles.

Microbeads: A tiny sphere of plastic (such as polyethylene or polypropylene) used especially as exfoliants in facial scrubs and body washes and to add texture to other personal care products. (Webster)

Microfibre: a fine usually soft polyester fibre having a diameter of less than 10 microns. (Webster)

Microplastic: Small pieces of plastic, less than 5 mm (0.2 inch) in length, that occur in the environment as a consequence of plastic pollution. (Encyclopedia Britannica)

Migration: The movement of a person or a group of persons, either across an international border, or within a State. It is a population movement, encompassing any kind of movement of people, whatever its length, composition and causes; it includes migration of refugees, displaced persons, economic migrants and persons moving for other purposes, including family reunification. (IPCC)

Minimalism: a style or technique that is characterized by extreme spareness and simplicity (Webster,) in this case, usage of products.

Mitigation (of climate change): A human intervention to reduce emissions or enhance the sinks of greenhouse gases. (IPCC)

Monoculture: The cultivation or growth of a single crop or organism especially on agricultural or forest land. (Webster)

Multinational: Having divisions in more than two countries. (Webster)

N:

Neurological: Anything concerning the structure, function, and diseases of the nervous system. (Webster)

Nitrous oxide (N_2O): One of the six greenhouse gases (GHGs) to be mitigated under the Kyoto Protocol. The main anthropogenic source of N_2O is agriculture (soil and animal manure management), but important contributions also come from sewage treatment, fossil fuel combustion, and chemical industrial processes. (IPCC)

Non-government organisation (NGO): A voluntary group of individuals or organizations, usually not affiliated with any government, that is formed to provide services or to advocate a public policy. (Encyclopedia Britannica)

Non-Renewable: Unable to be replaced or replenished once used. (Webster)

Nutrition: The sum of the processes by which an animal or plant takes in and utilizes food substances that enables its nourishment. (Webster)

O:

Ocean acidification (OA): Ocean acidification refers to a reduction in the pH of the ocean over an extended period, typically decades or longer, which is caused primarily by uptake of carbon dioxide (CO_2) from the atmosphere, but can also be caused by other chemical additions or subtractions from the ocean. (IPCC)

Organisation for Economic Co-operation and Development (OECD): The Organisation for Economic Co-operation and Development, abbreviated as OECD and based in Paris (FR), is an international organisation of 38 countries committed to democracy and the market economy.

Overfishing: To fish to the detriment of (a fishing ground) or to the depletion of (a kind of organism). (Webster)

Overpopulation: The condition of having a population so dense as to cause environmental deterioration, an impaired quality of life, or a population crash. (Webster)

Ozone (O_3): Ozone, the triatomic form of oxygen (O_3), is a gaseous atmospheric constituent. In the troposphere, it is created both naturally and by photochemical reactions involving gases resulting from human activities (smog). Tropospheric ozone acts as a greenhouse gas. (IPCC)

P:

Paris Agreement: The Paris Agreement under the United Nations Framework Convention on Climate Change (UNFCCC) was adopted on December 2015 in Paris, France, at the 21st session of the Conference of the Parties (COP) to the UNFCCC. The agreement, adopted by 196 Parties to the UNFCCC, entered into force on 4 November 2016 and as of May 2018 had 195 signatories and was ratified by 177 parties. One of the goals of the Paris Agreement is 'Holding the increase in the global average temperature to well below 2°C above pre-industrial levels and pursuing efforts to limit the temperature increase to 1.5°C above pre-industrial levels', recognising that this would significantly reduce the risks and impacts of climate change. Additionally, the Agreement aims to strengthen the ability of countries to deal with the impacts of climate change. (IPCC)

Pescatarian: A person who does not eat meat but does eat fish. (Oxford)

Particulate matter: The sum of all solid and liquid particles suspended in air many of which are hazardous. This complex mixture includes both organic and inorganic particles, such as dust, pollen, soot, smoke and liquid droplets. These particles vary greatly in size, composition and origin. (WHO)

pH: pH is a dimensionless measure of the acidity of a solution given by its concentration of hydrogen ions ([H+]). (IPCC)

Photosynthesis: The process by which green plants and certain other organisms transform light energy into chemical energy. (Encyclopedia Britannica)

Pollutants: Something that pollutes (Webster). In this case something that pollutes natural resources like air, water, land etc.

R:

Recycling: To reuse or make (a substance) available for reuse for biological activities through natural processes of biochemical degradation or modification. (Webster)

Renewable: Capable of being replaced by natural ecological cycles or sound management practices (Webster)

Retail market: The market for the sale of goods or services to consumers rather than producers or intermediaries. For example, a retail clothing store sells to people who will (most likely) wear the clothes. It does not include the sale of the clothes to other stores who will resell them. The retail market contrasts with the wholesale market.

S:

Supply chain: The chain of processes, businesses, etc. by which a commodity is produced and distributed: the companies, materials, and systems involved in manufacturing and delivering goods. (Webster)

Surface runoff: Surface runoff is the water that flows from rain over a land surface to finally reach a nearby stream. (Science Daily).

Sustainability: A dynamic process that guarantees the persistence of natural and human systems in an equitable manner. (IPCC)

Sustainable development (SD): Development that meets the needs of the present without compromising the ability of future generations to meet their own needs (WCED, 1987) and balances social, economic and environmental concerns. (IPCC)

Sustainable Development Goals (SDGs): The 17 global goals for development for all countries established by the United Nations through a participatory process and elaborated in the 2030 Agenda for Sustainable Development, including ending poverty and hunger; ensuring health and well-being, education, gender equality, clean water and energy, and decent work; building and ensuring resilient and sustainable infrastructure, cities and consumption; reducing inequalities; protecting land and water ecosystems; promoting peace, justice and partnerships; and taking urgent action on climate change. (IPCC)

T:

Telecommunications: Communication at a distance (as by telephone). (Webster)

The International Convention for the Prevention of Pollution from Ships (MARPOL): is the main international convention covering prevention of pollution of the marine environment by ships from operational or accidental causes.

Thrifting: Careful management especially of money (Webster). In this case it also means buying used clothes from thrift stores or flea markets at discounted prices.

Troposphere: The lowest, densest part of the earth's atmosphere in which most weather changes occur and temperature generally decreases rapidly with altitude and which extends from the earth's surface to the bottom of the stratosphere at about 7 miles (11 kilometres) high. (Webster)

U:

United Nations Framework Convention on Climate Change (UNFCCC): The UNFCCC was adopted in May 1992 and opened for signature at the 1992 Earth Summit in Rio de Janeiro. It entered into force in March 1994 and as of May 2018 had 197 parties (196 States and the European Union). The Convention's ultimate objective is the 'stabilisation of greenhouse gas concentrations in the atmosphere at a level that would prevent dangerous anthropogenic interference with the climate system.' (IPCC)

Upcycling: To recycle (something) in such a way that the resulting product is of a higher value than the original item : to create an object of greater value from (a discarded object of lesser value). (Webster)

Urban migration: The migration of people from rural to urban areas for better opportunities or jobs. (World Bank)

Urbanisation: The process by which large numbers of people become permanently concentrated in relatively small areas, forming cities. (Encyclopedia Britannica)

V:

VOCs—Volatile Organic Compounds: Volatile organic compounds, or VOCs, are gases that are emitted into the air from products or processes. Many VOCs are human-made chemicals that are used and produced in the manufacture of paints, pharmaceuticals, and refrigerants. (EPA)

Vulnerability: The propensity or predisposition to be adversely affected. Vulnerability encompasses a variety of concepts and elements including sensitivity or susceptibility to harm and lack of capacity to cope and adapt. (IPCC)

Z:

Zoonotic: An infection or disease that is transmissible from animals to humans under natural conditions. (Webster).

Help Stop Greenwashing in India

---

greenwashing(n.)
The creation or propagation of an unfounded or misleading environmentalist image.

'100% natural', 'organic' and 'eco-friendly'.
These labels are easy to find, but hard to verify. How often do you choose a particular product over another just because it claims to be 'eco-friendly'? More often than not, these claims may prove to be untrue, or more so, half-true.

As brands rush to show that they are 'green', or in many cases 'greener than others' it is imperative that it be done with honesty and transparency. After all, well-intentioned citizens must not be misled into believing that the product they are buying is benefiting the environment, when in fact it may just be causing more harm.

I have started a petition on change.org urging the Government of India to develop regulatory guidelines to curb greenwashing in our country.

Visit this link below or scan the QR Code to join me!
Sign the petition to Stop Greenwashing in India.

change.org/StopGreenwashingIndia

Join the conversation on social media using these tags:
#StopGreenwashingIndia | @aakashranison